Inside Schools:
A Collaborative View

The Stanford Series on Education and Public Policy

General Editor: Professor Henry M. Levin, School of Education, Stanford University

The purpose of this series is to address major issues of educational policy as they affect and are affected by political, social and economic issues. It focusses on both the consequences of education for economic, political and social outcomes as well as the influences of the economics, political and social climate on education. It is particularly concerned with addressing major educational issues and challenges within this framework, and a special effort is made to evaluate the various educational alternatives on the policy agenda or to develop new ones that might address the educational challenges before us. All of the volumes are to be based upon original research and/or competent syntheses of the available research on a topic.

Inside Schools:
A Collaborative View

J. Myron Atkin
Donald Kennedy
Cynthia L. Patrick

 The Falmer Press

(A member of the Taylor & Francis Group)
New York • Philadelphia • London

UK	The Falmer Press, Falmer House, Barcombe, Lewes, East Sussex, BN8 5DL
USA	The Falmer Press, Taylor & Francis Inc., 242 Cherry Street, Philadelphia, PA 19106-1906

First published 1989

Library of Congress Cataloguing-in-Publication Data available on request

British Library Cataloguing in Publication Data

Atkin, J. Myron
  Inside schools: a collaborative view. — (The Stanford series on education and public policy).
  1. United States. Schools. Cooperation with universities. 2. United States. Universities. Cooperation with schools
  I. Title   II. Kennedy, Donald   III. Patrick, Cynthia L.
  IV. Series
  371'.00973

  ISBN 1-85000-432 3
  ISBN 1-85000- 433 1 Pbk

Printed in Great Britain by
Redwood Burn Limited, Trowbridge, Wiltshire
and bound by Pegasus Bookbinding, Melksham, Wiltshire

Contents

Acknowledgements

The picture of teachers, students and schools presented in this volume is drawn from the work of scores of people who participated in a multi-year program of collaborative research in six school districts. The perspectives, however, are those of the authors alone. We believe our views are consistent with the observations and conclusions of our colleagues in the project, but we have drawn on only a small portion of their total work. Any of the other participants probably would have chosen different facts and themes for emphasis.

The professors, teachers, research assistants, school administrators, advisors and support staff whose work provided the basis for this book are listed as in the appendix, as are the supporting philanthropic foundations and public school districts. Without their collective efforts, this volume could not exist. We want, however, to highlight the special contributions of two people.

From the outset, Mary Nur, who served as Assistant Director of the project, was a critical figure in forging the collaborative activities that characterized our work. She not only established contacts between those at the University and those in the public schools, she sustained them on a daily basis by her special skills and her deep knowledge of the characteristics, strengths and weaknesses of both types of institutions.

Bob Calfee was the Associate Director of the project, a role he assumed after the work had begun because he and the Directors saw the need for greater conceptual unity in a large and decentralized venture conducted by dozens of independent and forceful people. He is responsible in no small measure for whatever cohesiveness is apparent in this volume, or in the other reports that eventuated from the study. Of course, neither he — nor anyone else but the authors — bears responsibility for the selection of facts and the basic viewpoints that are presented in the following pages.

J. Myron Atkin
Donald Kennedy
Cynthia L. Patrick
August 1988

Chapter 1

Schools, Universities and the Origins of the Study

This is the story of how a university committed to the advancement of scholarly research and six neighboring public-school districts joined forces to address serious educational problems. It is a local story, special and particular; yet the reasons for the project, the hopes and accomplishments of the participants, the guiding premises of the collaboration, and the problems that arose may offer a case study from which others can profit.

But a case study of what? What general conditions, issues and challenges do the activities reported here attempt to address? Why might others benefit from this story?

In 1981, when the project was conceived, joint efforts between public schools and research universities were rare. Although the two types of institutions have had profound influences on one another ever since both were established, they seldom have been involved in cooperative and serious examination of their relationships and common aims, and hardly ever in joint enterprises that might affect each other's programs. Historically, faculty and administrators in universities have seen themselves as worlds apart from those in the schools, and high school teachers have seldom been in a position to offer suggestions about college-level programs (though they have always understood, sometimes painfully, how their work is shaped by decisions at the universities). In fact, for virtually all of this century and much of the last, the contacts that did exist between high school and college teachers tended to be tinged with friction. In particular, many high school teachers were irritated and sometimes resentful of the influence of university admission policies on the topics they had to teach and, sometimes, on the ways in which they taught them.

Adding to the friction and fueling the resentment, it had become routine over the decades, and even expected, for university professors to complain about the quality of the programs young people receive in high school, usually blaming the teachers; poor preparation in high schools it was claimed, prevented professors from building an educational program around challenging and important content.

A large, influential and visible (but possibly transitory) departure from this state of affairs surfaced with the curriculum development activities that were initiated in science and mathematics just before and for about ten years following the launching of Sputnik I, the mid-1950s to the late-1960s. For a relatively brief and unusual period, outstanding university-level scholars became deeply committed to and visibly involved in projects designed to shape curriculum and improve teacher education for pre-collegiate institutions. Though these activities were not often characterized by significant collaboration with faculty and administrators in high schools and elementary schools, and certainly by nothing like parity in their influence on the resulting curriculum, professors were not entirely cavalier towards teachers — at least in several of the projects — and relationships between university-level teachers and those in the high schools became somewhat more collegial. More important for the present case, the memory of that period, and perhaps its myths, shaped some of the motivation of participants in the Stanford project, and also its goals.

For accuracy and greater completeness, it should be noted that even before the 1950s and 1960s, antagonism between college and high-school teachers was not universal. At almost every university, there were a handful of professors who met regularly with high-school teachers, and sometimes with students, to ease problems of transition between secondary school and college. Some high-school teachers had become particularly knowledgeable and influential about such issues as they helped to design courses and write texts that articulated well with college-level programs. But relatively smooth and constructive contacts were limited, at least in the proportion of people who might have become engaged in such activity. Cultivation of more productive relationships between high schools and colleges was a marginal activity at both types of institution.

Noteworthy, also, is the fact that while criticism by university professors of the secondary schools is a time-honored ritual in the United States (much more so, in fact, than in any other Western country), the calls for change from the academy acquired a markedly different tone after World War II. Prior to 1940, it was not uncommon for a professor to have begun his career as a teacher in high school. From the new position, the professor might (and often did) comment freely about the shortcomings of secondary education (sometimes to confirm his new status) but the criticisms were frequently well-informed since they were bred of first-hand knowledge; the comments were often credible even when they hurt. After World War II, however, with the sudden expansion of higher education and the accompanying urgent need for college-level instructors, a much smaller proportion of college-level teachers than before were drawn from high schools. The new teachers joined higher-education faculties immediately on receiving their advanced degrees.

Another significant factor in shaping post-war relations between universities and schools was the sharp shift in the make-up of the population of students enrolled in high schools. Up to 1939, about three-quarters of those who entered high school dropped out before completing four years of work, and only a relatively small proportion of the remainder went on to college. University teachers during the first four decades of the century, and earlier of course, met a highly selected group of young people in their classes. High schools were considered to be college-preparatory institutions by most of those who attended or taught in them. For the most part, they served an intellectual elite. Virtually everyone in such schools subscribed to the same goals. There was homogeneity, in fact, not only in outlook, but in race and, to a lesser degree, in social class.

Quite a different group of students started to complete high school *and* go on to college by the 1970s. By 1979, about 75 per cent of nineteen-year-olds were receiving a high-school diploma, compared with about 25 per cent before the War, and about 50 per cent of the graduating group were moving on to some form of higher education. During a relatively short span of years, American secondary education had become almost as universal as American elementary education. College professors began to see large numbers of students quite different from those who, like them, had graduated from high school a decade or two earlier. It was a population with which they previously had little association. Criticisms from the academy about the quality of secondary schools, therefore, were both less informed about the dynamics of secondary-school teaching because fewer college teachers had worked with the new population of students, and at the same time less comprehending of the new missions that these institutions had acquired or had thrust upon them with the broadening of their demographic base.

There were other factors associated with the widening rift between the academy and the high schools. As stated, after World War II, it became much more common to begin teaching at a university immediately upon graduating from one. But accompanying the rapid change in background (and age) of the professoriate in the late 1940s and the 1950s, the academic expectations of professors changed, too — in response to a shift in the mission of universities.

Specifically, the research component of university-level activity rose enormously in scale, prestige, and influence during the War. Colleges had been seen primarily as teaching institutions before 1940. By 1945, a new and deep awareness of the practical importance of academic research had arisen. Radar and rockets had made their impression on the public during a period of extraordinary crisis. Most strikingly, the atom bomb, which had led to a rapid and decisive end to the conflict, brought people to a new respect for the impact on world events of researchers who had seemed in earlier decades to be preoccupied with impractical, remote and largely incomprehensible activities. Einstein became something of a folk hero. Funds were funneled to universities to do more research. The National Science Foundation was created in 1950 to accentuate this development. In the process, and without apparent forethought, the teaching function of universities, the element that most strongly linked high-school teacher with university instructor, began to recede in importance.

Furthermore, scholars, particularly scientists, were people to be taken seriously, not only in their specialties, but in world affairs. The stage was set for professors as a group to become much more influential outside the academy than they ever had been before. If they should choose to use this new power to act on their convictions about the proper focus of secondary education, and gradually many of them did, they were a force that commanded respect and attention. With support primarily from the National Science Foundation, university professors began to develop programs designed to reshape the secondary school curriculum in science and mathematics.

While secondary-school teachers and administrators did not always welcome the ardent attention they began to receive from universities, those inside the high schools seemed more receptive to change than one would ordinarily expect when one's own institution, particularly a well-entrenched one, is the target of externally generated pressures for 'reform'. Teachers, too, had been impressed by what scientists had accomplished during a national crisis. Furthermore, they were flattered by the attention. It was being universally affirmed by some of the nation's most prestigious communities, on a scale unseen before, that teaching high-school students is a socially important task. The nation's future is at stake.

Of consequence in understanding the history of relationships between universities and schools is the subtle matter of teacher identity. If a high-school teacher in 1960 were asked to state his occupation, he would, more often than not, respond with 'biologist', or 'historian', or 'mathematician'. By 1980, the answer became 'teacher'. The reasons are probably strongly associated with the changed student population, the consequent broader educational mission of high schools, and also to the rapid unionization of the profession. But this sense of personal indentification of the high-school teacher with the academician, so prevalent before the rapid shift in the demography of the high school, was probably a powerful factor in the receptivity of the schools to the attempts by people in the university community to produce change during the 1950s and 1960s.

The joint effort described here that came to be called 'The Study of Stanford and the Schools' is a case, then, of representatives of two major types of educational institutions, still largely isolated from the other and still mutually suspicious as a result of frequently hostile interactions, acknowledging anew some of their common problems, recognizing the seriousness with which the public was taking the matter of educational quality, and taking the first, tentative steps toward trying to work together in a new way, profiting, it was hoped, from what had been learned during previous periods of cooperative activity.

The collaborative effort was conceived in 1981. Work began in 1982. The rising national concern about, and criticism of, seconday schools this time seemed largely to be the result of America's perception of its declining position in world economic competitiveness. Fervor for reform was high and soon to crest. Changes were demanded, urgently. Yet, in view of the ambivalent history of associations between universities and schools, it was an unexpected alliance between Stanford and the nearby districts.

Teachers and school administrators operate in a huge enterprise. There is an incessant demand to meet the needs of each child, the expectations of their parents and the requirements of the nation — all at a high level of quality, day-by-day and at the lowest possible cost. Research universities have their own pressures; however, professors at these institutions train a relatively small number of those who staff the public school system, and university faculty are encouraged to take the long view. To the extent that universities focus on pre-collegiate education at all (other than as a source of entering students), they engage in scholarly activity that questions basic assumptions about schools and schooling, analyzes and criticizes current practice, and develops inventive new approaches to the education of youngsters.

Nevertheless, the 'Study of Stanford and the Schools' was launched. It was to be research-oriented, data-based, reflective — yet geared to action. The aims were both to understand the schools better and to employ a collaborative process — involving teachers, school administrators, professors and graduate students — to reach that understanding and make things better.

There was more-than-a-little initial skepticism, particularly on the part of some school administrators and teachers. The risk of unmet expectations was not negligible. But against a background of a perceived national educational crisis and local good intentions, the alliance was formed. Commitments were made by all parties. Funds were secured from outside and inside sources, and the work began.

What common interests were identified? How were they nurtured? What new ones emerged? What talents did each group of participants bring to the Study? What base was established for ongoing work? What problems loomed large as the new relationships took root? The answers to those questions are one theme of this monograph. The other themes relate more directly to the state of secondary education today: complex, conflict-ridden, often impoverished, and increasingly subject to regulation. These issues are examined in concrete and operational terms. How do teachers and school administrators approach their day-to-day challenges to sustain and improve the quality of the education they provide — and how are those responses modified as a result of assistance from a nearby university?

Chapter 2

Launching the Study

The Study of Stanford and the Schools started with several different motivations, but a single aim: to make a serious contribution to our understanding of what really goes on in public secondary schools — not what is supposed to go on, but what actually happens. The underlying motivations included at least the following: a desire for Stanford to re-establish functional working relationships with neighboring school districts, for their benefit as well as Stanford's; to push work in and with the schools to a higher place on the School of Education's agenda; and to interest a wider range of Stanford faculty, outside the School of Education, in the important matters of schools and schooling.

So we started, as university people are apt to do, with a proposed collaboration centered on research to find out what the schools are like today. It had a half-dozen subprojects in the beginning: the 'real' curriculum, the impact of outside agencies on secondary-school programs and organization, the development of personal responsibility in students and what the schools should do about it, the relationship between testing and teaching, the impact of technology on students and the classroom, and teacher education. Later, a project on global education was added, called 'American Schools and the World'.

These were, naturally enough, reshaped and restructured somewhat as the Study went on. But once we began in earnest, something very much deeper and more significant than mere topical restructuring began to take place. Although the original research objectives and those that were added later continued to provide the agenda for investigations, the collaboration itself — that is, the collaboration between Stanford faculty and graduate students on the one hand and superintendents, principals and teachers in the neighboring schools districts on the other — became not solely an instrument *of* the Study but one of the central issues *for* the Study.

An unexpected, gradual and partial transformation of form into substance, of method into subject matter, took place, so that the collaboration itself became one of its own most important outcomes — thereby assuring the continuation of the Study of Stanford and the Schools. What had begun as an episode of collaboration to understand high schools and how they might be improved had been converted into what will be, we now believe, an enduring relationship. How?

Schools, like other institutions in American society, have grown enormously more complex, both as institutions and in terms of the expectations society places upon them. In addition to a dramatically more diverse set of students, both in ability and expectations, schools today reflect a more complex structure and often more prescriptive environment than in past decades. For administrators in the schools, it means a multiplied array of influences: from parents and parent–teacher organizations, from district boards, from administrators at the district level, and increasingly from the state government. Not only is their accountability broader in range; it is couched in terms of objectives that are increasingly heterogeneous, almost bewilderingly so. These changes have all made schools more difficult places to work in and to understand.

As much as anything else, it is this complexity of context, often of interest to professors, that made the relationship between Stanford and the schools attractive from the schools' side, as well. As a graduate research assistant put it, 'The people who work with this mess are really hungry for solutions'. Their hunger was matched by that of the Stanford researchers, but of course there was, and remains, some difference in objectives. This led to an inevitable tension between the researchers' need to simplify questions and situations in order to obtain clear answers, and the teachers' need to hold on to reality by retaining its full complexity. Teachers don't operate in simplified worlds.

Although those differences were significant, the power of the need to understand was greater. And as time went on, it became clearer that the differences could only be worked out by working closely together, struggling for solutions, and finally reaching agreement on what to do. Not only did those in the schools want to be co-investigators rather than subjects; the success of the venture required it. That was one major force in the transformation of process into substance.

Another had much to do with recent history. In any early-1960s summer at Stanford, school teachers were the most abundant form of life. Various masters' degree programs with in-service components attracted hundreds of teachers for the summer quarter. Paul Hurd, Stanford's distinguished professor of science education, would dragoon members of the biology faculty into holding special seminars or supervising directed reading projects for the students — and the faculty members ususally thanked him later, because their summer students often turned out to be even more engaged and interested than the ones who were being taught during the other three quarters. Thus, at least in the sciences, there was a rich commerce between the secondary and post-secondary sectors.

But by the early-1980s, the academic programs devoted to teacher preparation in Stanford's School of Education had been reduced to one, with an enrollment of fewer than thirty students (now revitalized and expanded, fortunately). Summers had become quiet in the School of Education. There was no significant federal support for either pre-service or in-service education of secondary-school teachers in the 1970s and into the 1980s, and in-service summer programs leading to the Master of Arts in Teaching, of the kind that filled campuses with the bright summer migrants of the early-1960s, are no longer a prominent part of the scene. Today's teachers, even the younger ones, are aware that something is missing, and it would be surprising if they did not yearn to replace it. That desire for an expanded and more intense collegiality, we would argue, has been another force for stimulating collaboration.

But in its initial intentions and design, the project looked very much less collaborative than it eventually turned out to be. Both of the institutions involved — Stanford, and the schools— had something to gain from mutual involvement. The schools saw a real chance to learn some things that could illuminate their own decisions, and perceived that affiliation with a leading university might provide them with a powerful political partner. Stanford felt it had lost some of its involvement with the professional subject matter of a professional school, and needed to re-establish it; and it also wanted to improve the continuity between its School of Education and the rest of its own faculty.

So to both parties, the project initially represented an expansion of Stanford's research agenda; Stanford would 'go into' the schools, investigate, and proffer its findings. This would be useful to the schools. It would provide possible material for school improvement elsewhere, would contribute a data-based analysis of school performance to the national debate about the quality of schooling, and would perhaps provide affiliations useful for the future.

But in the end, this last outcome became the most significant one. The process of conversion began slowly, almost imperceptibly. Formal cooperation came relatively easily and relationships were established with six schools districts in the region. General objectives for the study crystallized, and participants agreed to them. But there was, in the beginning, some frustration on the part of the school administrators. They saw a need to get started but found some lack of clarity in Stanford's original intentions. In part, that impression of vagueness was justified. Spurred by the need to get involved and undertake some serious analysis, the university researchers had formulated plans that were in some cases incomplete, and in others somewhat unrealistic. The context in the schools had changed markedly over the past two decades, a period during which 'on the ground' involvement by university researchers had decreased sharply. On the other side, the principals and superintendents brought with them some expectations that had more to do with practical advice than with research. There was disappointment, for example, that Stanford scientists outside the School of Education were not going to work on the *content* of curricula in the Study of Stanford and the Schools. Why, one administrator wanted to know, were not Stanford people 'identifying the most important ideas to be taught in a particular discipline?'

Though Stanford took the lead in designing and launching the Study, it had to promise to be worth the effort to all the participants. What further considerations led the University to suggest a close collaborative activity with nearby schools? There were no irresistible pressures for such an initiative, inside or outside the University. Research, after all, is the priority of a research university. To be well-regarded, professors, even education professors, need not aim their scholarly activity toward the resolution of practical problems in the schools, and certainly not in close collaboration with school teachers and administrators.

Intellectual productivity at the university level has its own scholarly rules and forms that evolve within the disciplines; professors can continue to do work that will be well-received within their fields by following well-accepted canons of academic inquiry. Such norms tend to place high value on 'fundamental' research, which is considered to reflect the ability to think abstractly, a quality much admired in the university community. Most university professors will assert that general concepts, relatively abstract formulations, are also those with greatest applicability; by definition, they apply to the greatest number of cases. These academic norms that value highly generalizable results characterize not only the field of education but other professional fields, too.

Outside the university, there is ambivalence about its internal values. On the one hand, there is in American universities a 'land-grant' tradition dating from the middle of the nineteenth century that emphasizes the practical applications of intellectual inquiry. True, such a tradition is associated most closely with the major public universities of the Midwest, and to a significant degree with almost all the state universities that were created afterward. Nevertheless, Stanford, as much as any private university in the country, reflects a similar set of origins and values. Leland Stanford wanted a practical place when he and his wife created a university to memorialize their son. He also chose as the first president David Starr Jordan, a man whose academic background had been at Indiana University (where he had gone directly from serving as a high-school zoology teacher in Indianapolis), one of the original and great land-grants, and Cornell. The latter institution at the time of Stanford's establishment had more of a public-service orientation than almost any other private university in the East because a significant part of its funding came from tax dollars. New York, unlike the Midwestern states, had a powerful and entrenched set of private-university interests determined to block establishment of a separate state university that would compete with places like Cornell, Syracuse and Columbia. Therefore, public functions identified with the land-grant concept were attached for budgetary purposes to private universities: agriculture at Cornell, for example, and forestry at Syracuse.

On the other hand, the public admires distinction. If Stanford chose to earn fame from its contributions to theory, and was successful, then those who cared about the place took understandable pride in the accomplishments, even if they were only imperfectly understood and had little apparent effect on practical affairs.

The Study was to focus on the practical business of improving education in the more-or-less short term, an aim that was not prominent in the School of Education at the time. It could be seen as somewhat risky, as well as unusual, for professors to take such a step; researchers are accustomed to subjecting their work to the judgment of peers. That expectation and responsibility carry their own sets of apprehensions. But to do something as public as the Study promised to be — and to include as primary audiences the public, teachers and school administrators — meant engaging in an activity with no set of tested procedures or clear criteria for determining success. What, then, were the hopes at Stanford that counter-balanced the risks? What might professors, students and the institution gain?

Professional Roots

The answer to those questions are almost as varied as the number of University participants in the Study. One of us, Atkin, was the new Dean of Stanford's School of Education at the time. He had come to Stanford in 1979 committed to emphasizing its professional roots. He said as much when taking up the position, even characterizing himself as an 'educationist', a label meant to emphasize his identification with those who staff the schools. His career pattern before coming to Stanford underscored the point; he had taught science for two years at the high school level in New York City and for five in elementary schools. His formal training had been in chemistry at the undergraduate level and science education at the graduate level.

On moving to a university appointment, he attempted from the outset to test the relationship between university-based scholarship and the practical problems of improving the quality of education at the site where it is provided to young people. His first major professional activity after arriving at the University of Illinois in the mid-1950s was to co-direct a curriculum project in astronomy that represented one of the first attempts by the National Science Foundation to modify the curriculum (and to a degree, teacher education) at the pre-high school level.

Atkin was mindful of the fact that the Stanford School of Education's origins could be traced back to the creation of the University, when Jordan brought Ellwood P. Cubberley to the West from Indiana to build a small department of education for the purpose of exerting a positive influence on high schools — particularly those that sent students to Stanford. In the 1980s, it is easy to forget that as recently as the pre-World War II period, even the most outstanding universities did not attract all the students of the quality they wished to have. Education departments were established near the turn of the century at institutions that emphasized research and scholarship to assure a flow of suitably qualified candidates to higher education. This was no less true for private institutions than for the major state universities created by the Morrill Act of 1862. (Education departments at the state-supported 'colleges', the former 'normal' schools, had quite different origins.)

In the intervening decades, both before and after World War II, education units at research universities became schools and colleges instead of departments and began to stress academic scholarship. Along the way, they changed their focus markedly from direct involvement in improving schools to conducting research about them, assuming that the results of the research would find their way into classroom practice. Faculties at highly regarded schools and colleges of education tended to model their research on the social and behavioral sciences because modes of inquiry in these fields seemed to reflect the main priorities of the home institutions better than more practical work. At the most highly regarded schools, Stanford among them, priority in filling vacant positions went to hiring those with strong background in fields like anthropology, economics, political science and sociology — in addition to psychologists, who had long been prominent members of education-school faculties.

Universities *do* evolve as a group despite the very important differences among them, and the trend toward hiring social scientists at schools of education gained impressive momentum in the 1960s, with Stanford leading the way.

Even in educational psychology, educational history and educational philosophy — fields with significant roots in schools of education — research came to look more like research in psychology, philosophy and history as research goals and publication in scholarly journals were advanced more assertively at graduate-level universities. Scholarships that centered directly on schools as internally coherent institutions in their own right began to recede in favor of research and analysis that viewed educational phenomena in terms of economic, political, anthropological or sociological theory. In the process, university relationships with elementary and secondary schools became more indirect and tenuous. Many of the new professors, in fact, had never seen employment in schools, a marked change from the earlier decades of the century.

The new forms of scholarship were often first-rate, and the perspectives revealed by research with a strong social science base turned out to be important. To view schools, as many anthropologists do, as an institution for transmission of cultural values, for example, or to see them as 'loosely coupled' organizations that are difficult to standardize, as do many sociologists, are powerful perspectives for explaining many activities that otherwise seem puzzling or inconsistent. The point is that such professorial activity is often remote from the problems considered important by practicing teachers and school administrators.

The prestige of schools of education has never been high on university campuses. Many explanations have been offered. Harry Judge, Head of the Department of Educational Studies at the University of Oxford, who had been commissioned by the Ford Foundation in the late 1970s to study American graduate schools of education, suggested in his resulting monograph, *American Graduate Schools of Education: A View From Abroad*,[1] that the low status is attributable to the standing of the profession of teaching itself. Howe, in an introduction to the Judge volume, speculated that academic admission standards to schools of education are low: furthermore, as schools of education turned attention to the more inclusive group of students beginning to attend high school, a high percentage of whom would *not* go on to college, they moved toward a mission that university professors cared less about than college preparation.

Many school of education faculties, in moving toward the social and behavioral sciences and away from the schools, were also seeking to rectify their traditional low status on campus. With considerable forethought about the matter on the part of deans most active in this development, they were trying to look more like other departments on campus by producing research that would be accepted as competent work in the more-established and presumably more-respected disciplines.

In the process, the preparation of professionals to work in the schools was de-emphasized in the 1970s, particularly in schools of education at prestigious private universities like Chicago, Harvard and Stanford. Such moves seemed not to provoke deep controversy, either, inasmuch as there was a teacher 'surplus' that had resulted from the rapidly declining number of children in elementary and then in secondary school. Harvard dropped its teacher education program entirely. At Chicago and Stanford, the numbers of students annually entering teacher preparation programs fell to two or three dozen by the late-1970s — from a figure of well over one hundred in the early part of the decade. Duke dropped its education department entirely. To the extent that faculties at the most prestigious universities continued to view the training of practitioners as an important function of schools of education, they tended to emphasize the more-indirect goal of preparing people to teach at *other* universities where teachers presumably would be prepared in larger numbers.

Gradually, however, an increasing number of education-school graduates took up employment not in universities where jobs were drying up, but in policy-making bodies, such as staff to legislative committees, or in contract-research firms, where they could use more directly the skills that had been the focus of their graduate programs. They became further removed from classrooms, compared to education-school graduates of a decade or two earlier.

Just as Atkin was preparing to move to Stanford in 1979 to assume the Deanship, a faculty committee was considering the possibility of recommending elimination of the Stanford Teacher Education Program (STEP). This set of deliberations was the logical extension of a long series of actions within the School during the preceding fifteen years in which subject-matter specialists on the faculty in fields like science and mathematics, who taught the most-clearly relevant courses for teachers and prospective teachers, were gradually replaced, as they retired, by social scientists. Atkin resisted termination of the teacher-education program as one of his first acts, and during the 1979–80 academic year, the faculty accepted continuation of the STEP. Perhaps the faculty sensed that elimination of the program would represent a point of no return. It was less obvious, however, what might be done to relate research more directly to educational problems as the problems were understood by teachers, school administrators and the public.

A New Round of 'Reform'

In 1980 and 1981, it was becoming increasingly obvious that fresh attention was being focused on pre-collegiate educational institutions by the public. Studies about high schools had been initiated by the Carnegie Foundation and the Carnegie Corporation. The Secretary of Education was about to ask a national commission to make recommendations concerning how schools could be improved. The stirrings were significant, and the news media were beginning to turn impressive attention to the matter. The emphasis in the public discussion was on the shortcomings of the schools. The CBS television network broadcast a three-hour special titled 'What's Wrong With Our Schools?' anchored by no less a figure than Walter Cronkhite, almost universally considered the most-credible figure in television broadcasting, and one who inspired deep trust among his viewers. There were weekly newspaper features and editorials lamenting a perceived decline in standards and a precipitous drop in the academic ability of those choosing teaching as a career. The national mood was not merely one of sensible concern; an atmosphere of crisis was beginning to develop.

The public attention-span in the United States usually is not long; the media spotlight seems to shift rapidly. Pre-collegiate education had received unaccustomed attention in the late-1950s after the launching of Sputnik I. The schools were an important part of President Johnson's Great Society Program of the mid-1960s (in fact, Johnson wanted to be remembered as the 'Education President' before he was derailed by Vietnam), but this time the attention was on poor children and racial minorities rather than on the most talented young-sters, as was the case in the Sputnik era.

In the 1970s, schools receded somewhat from media attention, though there was a steady stream of conflicting comment about 'joylessness' in the classroom, about schools as an instrument to replicate an oppressive social structure, about children not being able to read because they were not taught by phonetic methods, and about the need for renewed attention to the 'basics' (usually defined as the three Rs). A key feature of popular education discussion in the 1970s, as always part of a much larger picture, was a general disenchant-ment in the public mind with the power of government to intervene significantly to alter the life chances of its citizens. The general theme, if one was clear, was that government should retreat from most social programs. Americans were told about the failure of legislative attempts to eradicate poverty, improve the penal system, address housing problems or provide effective health services. For about ten years, a stream of newspaper articles and television reports told of public housing programs that had destroyed a sense of community, about prison reforms that seemed to be associated with increases in recidivism, about welfare programs that seemed to perpetuate poverty. Presidential candidates — most pointedly Carter and Reagan — capitalized on and strengthened national skepticism about government by running against Washington and the 'bureaucracy'.

Most political leaders tapped the anti-government mood and took steps to reduce local taxes, much of which was collected for schools. This move was made politically more palatable by the fact that the number of children in schools was declining, and domestic expenditure was shifting from the young toward the elderly.

The sudden flush of attention to schools in the early-1980s engendered considerable nervousness among teachers and school administrators. Would they, once again, be blamed for educational ills? However, some educational leaders — a small number of teacher-union officials and some state school superintendents of schools, for example — saw the period, far-sightedly as it turned out, as one of opportunity to redirect not only national attention but public and private resources toward the improvement of education. Coura-geously (after all, the potential for criticism of teachers and school adminis-trators was great), they gradually began to embrace the rising pressure for reform without knowing quite what it would bring.

New Priorities

At Stanford, Atkin saw the new national mood as offering a chance to further enhance Stanford's programs in education by harnessing some university efforts for an attempt to ameliorate serious problems. He convened a group of professors in the School of Education to consider whether or not it would be beneficial for the University to initiate its own examination of the American high school. Eight members of the faculty met over a period of several months. The general idea seemed attractive. During the course of the discussions, broad outlines began to emerge for a major study.

A central issue for the faculty was how to plan an activity that would fit Stanford's recent history and evolving priorities. What might characterize a Stanford contribution, as distinguished from university-initiated activities that had arisen or were being discussed at other campuses? What particular contribution might Stanford make as the educational enterprise was readying itself for another round of 'reform'?

First, the planning group knew of only one initiative under way or in the planning stage that was likely to be based on original inquiry. Most seemed to promise informed opinion, probably considerable insight, but not research. A data-based investigation that built upon Stanford's strengths would offer a different, and complementary, contribution to the national debate. It could also, in projecting Stanford prominently into a relationship with the schools, heighten public knowledge and appreciation of the connection between scholarship and improved practice — a goal that the Stanford faculty considered to be of highest priority, even if it did not know clearly how to achieve it.

Second, Atkin and several other faculty members had a deep suspicion of the mischief often produced by acting on global analyses of problems in education. Most reform reports seem to talk about all schools and address their recommendations to everyone. Such perspectives tend to be inattentive to and unappreciative of particularity. They seem insensitive to the extraordinary variation in American schools. As a result, they breed global 'solutions', often poorly adaptive in local circumstances. As much as possible, the work at Stanford, the planning group thought, should be finer-grained, reflecting viewpoints developed at the level of classroom, school and district. None of the other studies under way or, as far as we then knew, being planned, promised to highlight a bottom-up view of the problems of schooling.

Third, Stanford did not want to produce yet another study that would result in attacks on teachers and administrators. Insofar as possible — convinced that schools are staffed by well-intentioned and skillful people, no more characterized by stupidity, laziness or capriciousness than those in other occupations — the planning group wanted at a minimum to reflect the potential of those who staff the schools to take the lead in addressing serious problems, when aided by a university. Gradually, we articulated another important goal: to produce an antidote to the reports that were expected to be released in the coming month and years by presenting a sympathetic portrayal of the ability of teachers and administrators to deal intelligently with the difficult problems they face, when fortified by the improved data-gathering and analytic capability that the University could provide.

A fourth, and related, point: many education faculty members were beginning to become sensitive to the trend toward increased centralization of power in the formulation of education policy. State education departments were becoming more assertive everywhere. A degree of demoralization was beginning to surface at the local level where principals, teachers and school superintendents were starting to voice concerns about loss of autonomy, not only because the shift of decision-making authority to the state level restricted their own range of choices, but because they often felt that remedies developed in the state capitol were insufficiently sensitive to local aims and needs.

Finally, Atkin was committed to try to modify some of the intellectual life of the School of Education to encourage research that would have short- and medium-term influence on practice, not only during the course of the Study, but afterwards. While several professors in the School worked closely with teachers and administrators, more often than not the studies they conducted were *about* the schools; that is, children, teachers and school organization were the 'subjects' of research. While such investigations often contribute importantly to theory and can produce perspectives that are useful in thinking about schools and teaching, they are not always seen by people in the schools as useful. Almost always, the professors who conduct the research believe there are important applications for their discoveries, but they seldom work with teachers and school administrators to find out what those applications might be. There is often the unstated assumption that the implications of the research should be worked out by the 'consumers'.

Such splits between theory and practice are by no means new, and they will always exist. Professors gain status by insightful and original contributions to theory; it would be destructive at a research university, even for a professional school in such a place, to attempt to direct all of its activities toward the amelioration of practical problems. Nevertheless, Atkin felt that intellectual styles at universities might be changed for the better if a somewhat greater share of university-based scholarship could be illuminated by the difficulties faced by those who have legal and professional responsibility for running the schools. He did not see the developing Study as an attempt to reorient Stanford toward a service mission, though this might be an attractive side-effect. Rather he thought professional and student scholarship itself might be enhanced by an attempt to cope more directly with the complex and untidy world of the classroom.

More fundamentally, for the Study to have lasting effect at Stanford, it would have to be seen by the professors themselves as aiding their research. Public service can be a powerful incentive if the perceived crisis is sharp enough. But to affect a university more profoundly, students and professors must be captured intellectually by the challenges associated with a new type of work.

Partly as a consequence, Atkin decided to enlist School of Education professors by inviting them to join the Study if they saw it as a potential vehicle for pursuing their own research interests. He established three conditions, however: Professors must be willing for the work to be collaborative with people in the schools. They must consent to cooperate with other professors at Stanford. And the research proposed should have the potential for recognizable impact on the schools participating in the Study over the short- to medium-term.

Kennedy's Concerns

Donald Kennedy was installed as Stanford's President in September 1980; very shortly afterward, he began to voice his views about the importance of public service as a part of Stanford's mission and as a serious commitment of more of Stanford's students. For him to join enthusiastically in the Schools Study was a natural extension of his expressed priorities. But eventually he became a co-principal investigator — a highly unusual step for a university president, and all the more striking because the new study was in the field of pre-collegiate education, hardly a traditional priority of a research university. He said, in 1980, 'Only if the major research universities care about elementary and secondary schools, and about their own schools of education, will others take these institutions with the seriousness they deserve'.

Some of Kennedy's reasons for becoming involved in the Study were identical with Atkin's, whereas others were different. The motivation was most similar where the direction and purpose of the School of Education were concerned. From the presidential perspective, it seemed strange to see a professional school so conspicuously alienated from the profession for which it was named and for which, presumably, it existed. As a member of the biology faculty twenty years earlier, Kennedy remembered a much richer intellectual commerce between the School and the teaching profession, and had participated in summer-quarter seminars for in-service science teachers seeking advanced degrees. To find the interaction rate between the Stanford School of Education and the public schools so low thus provided sharp historical contrast. It was as though the Law School had somehow become disconnected from the practice of law, or as though the Medical School had decided to train only medical researchers.

The view from the President's Office was strange in another respect as well. If one were to rank major domestic policy challenges, one would surely have to put health and education on approximately the same level. Success in managing both is clearly critical to the well-being of our society. Indeed, if one were to rank the *investment* value of each kind of expenditure, that for education would rank higher since our social expenditures in this area are concentrated on the early part of the life cycle, whereas those for health are concentrated heavily at its end. These activities have traditionally had approximately equivalent claims on national investment, though the last decade's dramatic surge in health-care costs has altered that traditional equity.

It seemed that at Stanford quite a different balance of serious policy interest obtained. When the Study began in 1982, it would have been possible to identify several dozen Stanford faculty outside the School of Medicine who were seriously interested in pursuing, as part of their own research programs, problems related to health policy. But nothing like that number could be found outside the School of Education who would claim an equivalent interest in educational policy, though there certainly were some.

At the same time, Kennedy was becoming concerned about a different problem — one that could eventually be seen to merge with concern over the quality of schools and of schooling, and over the direction and mission of the School of Education.

This had to do with the choices undergraduates were making at places like Stanford about their own vocational directions and the role of public and voluntary service in their own lives. In the early-1980s, an external consensus about American college students had begun to crystallize; in belief, it was characterized by a conviction that the brightest and most able young people in the society were selfish and 'careerist', thoroughly committed to economic security and uninterested in the welfare of their fellow citizens. That consensus is aptly expressed in the following quotation, a journalist's account of the younger generation: 'Today's students are passive, conformist, and materialistic. They care about jobs, while the baby boomers (their predecessors of ten years ago) cared about life.'

That characterization seemed at odds with the motivation and the impulses of the students we saw around us. It ignored, furthermore, a set of national events that had served to diminish popular regard for public service. It had become a politically popular platform to demean the 'bureaucracy', and candidates for public office took turns doing so. The rising public criticism about the schools made it clear — if, indeed, it needed to be made clear — that teaching, as one of the service professions, was part and parcel of what was being decried.

Kennedy thus became convinced that the problem had to be worked at from both ends. Able college undergraduates needed to become convinced that 'public service', taken in its broadest sense, was a valued activity that could once again earn national esteem. At the same time, the institution had to develop more significant links with the schools, involve more faculty members and students in school-related research and thus develop the conviction that things could be made better.

Initiatives of the first sort have had some significant results at Stanford. The Public Service Center has now been established, and permanently housed. The Stanford-in-Government Programs now attract large numbers of undergraduates to summer and academic-year internship, in Washington, in Sacramento and with local governments. A Stanford Volunteer Network is now affiliated with the Public Service Center; it provides a contact point for community organizations wishing student volunteer assistants and also as a marshalling place for students who wish to volunteer. A major effort during each academic quarter sends over five hundred Stanford students out for a single day's 'outreach' service in the community. Endowed fellowships now make it possible for two dozen students in each summer, and three especially qualified graduates each post-baccalaureate year, to undertake internships or fellowships in the public and voluntary sectors.

These activities have brought a significant transformation in the attitude of Stanford students, and the Stanford community, about the merit and rewards of the service professions. Teaching is not specifically mentioned as a central objective; but tutoring programs and other activities that involve teaching are significant elements of the Stanford volunteer effort. And with the growing regard for public and voluntary service has come an increasing interest in teaching, seen in the increased migration of Stanford undergraduates to the Stanford Teacher Education Program and other teacher-education opportunities.

To Kennedy, the Stanford and the Schools Study represented that other part of the strategy. A formal research-based involvement with the schools in the region would, it was hoped, accomplish several purposes.

First, it would supply a signal of the importance attached by leading institutions of higher education to improvements in the nation's primary and secondary schools. In the 1960s and 1970s, save for the involvement of some university faculty in the secondary-school curriculum-reform movement, there was little institutional evidence in, or concern for, the public schools. The Stanford and the Schools Study represented a potential antidote for this history of neglect. A strong public commitment to such a venture might influence other leading institutions in the same direction, and provide a message to our own students about the value Stanford attached to the improvement of the educational system as a whole.

Second, there was the hope that a research effort of this kind might catalyze new involvements between Stanford faculty as a whole and those members of it who had their primary appointments in the School of Education. The problems of schools and schooling in America are complex and difficult ones, unlikely to be resolved without the thoughtful attention of a range of academic specialties — a range substantially broader than that to be found within the faculty of a single school. Thus, a significant objective was the forming of new collaborations, and the generation of new interest in educational problems among faculty outside the School of Education.

Third, Kennedy was eager to involve Stanford in a variety of joint ventures with external agencies that would, collectively, strengthen the University's engagement with public policy. There has been no shortage of Stanford involvement with national science or health policy; education, by contrast, is more diffuse, and more locally controlled. Engagement with the policy issues is possible only if there is a close practical contact with the enterprise itself, and the only way to establish that is by remaking the institution's connections with the schools.

The central objective, of course, was the research itself, and the understanding it might develop about how universities and their research objectives might be turned to the betterment of the educational system. But from the point of view of the institution as a whole, the other objectives were highly significant as well, and they served to attract the attention not only of the President, but of Trustees and other senior administrators to the Study of Stanford and the Schools.

The Local Schools Get Involved

If Stanford was to avoid global pronouncements, if its professors wanted to highlight the world as seen from the level of the school district and classroom, then superintendents, teachers, and school principals would have to be involved in the Study in a significant and different way, even centrally. For school-level professionals to play such a role, they would need assurances that the proposed work would not result solely in another burst of criticism of the schools and teachers, with a tendency to blame victims. Therefore, while professors were being invited to join the Study, discussions were initiated simultaneously with school superintendents about whether or not they would participate and under what terms.

Superintendents in nearby districts were invited to discuss the Study with Atkin and selected faculty of the School of Education. Proximity to Stanford was important in selecting school districts since a considerable amount of work in schools and classrooms was envisioned, necessitating frequent meetings of school district personnel and those from the University. Discussions were initiated first with superintendents of schools — rather than teachers or school board members — because the planners knew that no study could progress far without the cooperation of the chief educational administrator in the district.

Fortunately, there was a base of goodwill toward Stanford on the part of superintendents in the area that created a fertile ground for discussion. Several had studied at Stanford. Professors had worked in their schools. Stanford students preparing to become teachers had been interns in their districts. Probably as important as any other factor, there is a regional pride about Stanford University; affiliation with it is often prized.

To be sure, there were skeptics about the proposed arrangements. The principal of one high school urged the superintendent not to join the effort. 'We'll get ripped off,' he said. 'It will be another blast at schools, gaining publicity for the professor at the expense of teachers and school administrators.' It is understandable if teachers and school administrators are gun-shy about opening their doors to those who would report on what goes on inside. Journalists, as one group, have a tendency to dramatize the worst. Scholars from the university often receive well-publicized acclaim for deep criticism that sometimes seems narrow, offensive and occasionally biased in the view of those who are reading about their own alleged failings.

Teachers have been told during the past two decades, for example, that they stifle the creativity of children by the kinds of routines that they establish in classrooms. Worse, they have been told by some professors that they are the unwitting tools of a state wherein the function of schooling is to prepare passive workers for industry. In such criticism, there is seldom a view of how the teacher or school administrator sees the job, of success stories, or even adequate description of the conditions under which teachers work. Practical wisdom in the school is usually unappreciated. Teacher and administration morale is further depressed by what they often see as unwarranted attacks on their professional ability.

Nevertheless, the superintendents of those districts invited to participate, from San Francisco to San Jose and Milpitas, without exception, agreed to co-operate in the Study. They were assured that this Study would be different, and that they would be significantly involved in the formulation of recommend-ations that might result from analysis and interpretation of the data that would be collected. They were told that the Stanford faculty were interested in their substantive help. Furthermore, they were promised, this would not solely be a study of the schools, but also, intentionally, an examination of Stanford in its relationship to the schools. How might Stanford's teacher education program be modified as a result of the Study, for example? What kinds of research might be undertaken at Stanford that had not been undertaken before?

Several features of the Stanford proposal seemed attractive to prospective school-district participants. First, there was the opportunity to learn more about some of the severe problems the school districts were facing, and how they might be attacked jointly. Furthermore, Stanford tried to assure pros-pective participants from the districts that teachers and school administrators would be considered partners by the University professors rather than objects of study. Additionally there was the promise that professors and their students would be involved in developing concrete 'solutions' to educational 'problems' instead of solely publishing research reports usually impenetrable to non-researchers, and that a style of educational investigation would be developed that would survive any specific time-span for the Study; the co-directors would do their best to create the conditions necessary for a continuing process of bringing University resources to bear on the improvement of high schools.

Richard 'Pete' Mesa of the Milpitas School District said, 'We looked forward to gaining access to good people who could provide advice and counsel'. Paul Sakamoto, Superintendent of Mountain View/Los Altos:

> We are within close commuting distance of one of the major research institutions in the nation. I saw this as a great opportunity for the renewal of our staff. The notion of being able to experience the collaboration between Stanford faculty and our staff was very exciting. I hoped that the staff would be able to meet and work in person with a number of professors whose books and studies they had read. Another exciting concept of the Study was the opportunity for interdisciplinary collaboration. The idea that our teachers and administrators could work with professors representing several fields gave us the chance to be at the cutting edge of new knowledge and new approaches to sharing that knowledge.

Robert Palazzi, Principal of Aragon High School:

> We had been living unexamined lives. We desperately needed some assistance on targeted research — for example, demographic studies, student population, the insulation and isolation of an aging staff that has not entirely adjusted to the very changing population at Aragon from the population we had here in the 60s. Since most of the staff cannot afford to live in the district, many of them were not really aware of how different our school population is now from what it had been.

Or, simply, from Charlotte Krepismann, an English teacher at Mountain View High School: 'I liked the idea of a relationship with a major university'.

For both the school districts and the project directors, the goal that was minimally articulated at the outset of the Study later assumed prominence: to enhance local capability for addressing educational problems. As indicated earlier, there is, to some, a distressing tendency in American educational policy to make inappropriate decisions, even those that affect curriculum and teaching, at the level of the state capitol. In fact, there is a growing expectation that state legislatures will play a major role in initiating education reform.

Furthermore, with budgets tight at the local level, what modest data-gathering capability once existed is being severely taxed. Universities are good at data collection and theory generation. Those who staff the school system are good at interpretation of research results and developing lines of action to address educational problems. As the Study progressed, it became somewhat clearer to all participants that an important desirable outcome, if not featured prominently as a goal initially, was the improved ability at the local level to address important educational problems on a continuing basis.

Given these expectations, the superintendents and principals involved in the early planning stages were relatively tolerant of the topics chosen for study, which, essentially, were selected by professors. For example, if a university professor wanted to study the schools' responsibility for teaching certain goals associated with citizenship, personal responsibility, and health knowledge, the school administrators tended to indicate approval, even if the professor's topic might not have been at the head of the school administrators list of pressing problems. There seemed to be enough promise in the projected relationship with Stanford for people in the schools to go along, at least initially.

Note

1   Ford Foundation, 1982.

Chapter 3

Designing the Collaboration

How did working relationships develop between those from the University and those from the schools? Since the process of collaboration became an integral part of the substance of the Study, and one of its most satisfying and enduring results, what was there in the interactions between the schools and the University that led to such an outcome?

While it was affirmed repeatedly at the outset of the project that professors and those in the schools would work collaboratively, there was no clear conception at the beginning of what such collaboration might look like operationally, either in the schools or at Stanford. For those in the schools, there was willingness to proceed on the basis of declarations of good intentions, with feelings of possible misgivings largely unarticulated and certainly not prominent. It was rare for a University to make such a public declaration of intent to address practical problems; some teachers and school administrators undoubtedly wanted to be part of the excitement and to see up-close what would happen. Most of those at the University were just as unclear about the meaning of collaboration, but some of them, too, wanted to find out about a new, large, public and well-supported activity. Besides, graduate research assistants were made available to help with the work, a significant enticement for professors.

Confidence in a significant degree of collaboration may have dipped a bit in the school districts when it became clear that the professors were playing the major role in defining the issues for study, but the stated reason for this fact seemed at least initially acceptable to those from the schools. For professors to become deeply involved in an activity that promised to be unconventional, they at least had to be committed intellectually to the enterprise; the topics for study had to be seen by them as challenging and offering the possibility of yielding to the kinds of data collection and analysis toward which professors gravitate, which they often are very good at doing, and for which they earn prestige and promotions.

While the problems for research were selected initially by the professors, however, it became clear in several instances that the topics were considered as tentative by those from the University and readily modifiable by superintendents. Some professors were more responsive to the preferences of school-district collaborators than others; unsurprisingly, the greatest flexibility was evidenced by those professors who engaged in early and frequent discussion with superintendents, principals and teachers about the research that might be undertaken.

For example, a prominent theme in early thinking at Stanford about the Study was to examine the alternatives to the comprehensive high school in the light of developments since 1959, the year that James B. Conant's classic '*The American High School Today*' was published. However, when Michael Kirst, the professor who suggested this topic, and his colleagues, took the idea to school superintendents, they displayed no interest in re-examining the basic concept of the comprehensive high school. For them, comprehensive high schools was a non-issue, a tired and non-controversial topic. Superintendents were uninterested in alternatives to such schools; instead they wanted to know how to improve them. As a result, the University group began to concentrate on factors within the comprehensive high school structure that make it work or that present problems. One difficulty, for example, was the basic matter of course selection. If high schools are to serve all students, how do teachers and the administration assure that appropriate courses are offered and that students have access to them? As other faculty joined the group examining this problem, the research became somewhat more general: how do both external and internal pressures on the school affect youngsters, administrators and teachers? What happens locally when the state introduces a two-year mathematics requirement? What are the implications for local action of the rising number of children in the schools whose first language is not English?

Sanford Dornbusch, a sociology professor in Professor Michael Kirst's group, was interested initially in the flow and use of information within a high school, across a school district and between school districts and the California State Department of Education. He redirected the proposed work, however, when he learned that school principals were much more attracted to another of his interests, but one that he had used only as an example in initial discussions: How does family structure affect student effort, and what can teachers and school administrators do about it? Here's how Dornbusch described the shift:

As an educational researcher, I've done a lot of studies in and around schools, and like most old fogies, I've developed a way of doing things. I choose the topics to investigate, I promise and deliver feedback to the schools, and I hope the schools find my research useful. But a glass of wine with Sam Johnson changed all that.

At a get-together of practitioners and researchers working with the Study of Stanford and the Schools last year, Sam, who is principal of Capuchino High School in the San Mateo District, asked me what I was studying. Almost as an afterthought, I mentioned some analyses I had finished on the impact of family structure on school–family relations and on student performance.

'That's what you should be studying,' said Sam. He proceeded to tell me how the day-to-day issues at a high school reflect families' differing levels of participation and understanding of the work of the school. Families are unaware of ways they could help their children, and don't know how to translate their concern about education into action. Meanwhile, students are seeking to escape from parental controls, rejecting school as part of their revolt. I couldn't honestly disagree with him. From then on, I sought to include the study of school–family relations as a central theme of the Study of Stanford and the Schools.

One sub-group in the Study, it should be noted, the one investigating curriculum and teacher preparation in international dimensions of secondary schooling headed by Professor Hans Weiler, incorporated teachers as participating members of the working groups at the outset. 'We had them as part of a team. The idea was not to make them an external reference group', said Weiler. Members of this group, in fact, were unable when asked, to differentiate the specific influence of research assistants, professors and teachers.

In each of the areas finally identified for research, however, those in the schools and those in the universities came to feel the topic was important and worth the investment of the time that was required from all parties. In those cases where administrator and teacher preference did not figure noticeably in the selection of the topics for research, subsequent participation in analysis and interpretation of the data heightened not only involvement but personal commitment.

The purposefulness and sense of commitment were demonstrated, for example, by extraordinary attendance rates by school-district participants at

Study meetings, animated discussion and a large number of school-initiated suggestions for further examination of the topics under review. Superintendents and principals are busy people, yet in the few instances where a meeting had to be missed, unmistakable disappointment and regret were expressed. The opportunity to participate collegially in trying to figure out the implications of the data collected, in pointing out to university personnel how certain school practices might produce certain results, to have a chance to *think* about some issues in the schools without the immediate press of running them proved to be powerful motivators for teachers and school administrators.

Still, a distinctive, general feature of the early months of the Study was that initiatives tended to come from professors in making proposals for research. This point is underscored for two reasons: (1) It contrasts with joint efforts between universities and public schools in which the declared initial objective of the collaboration is to provide assistance to school districts in the amelioration of problems the districts identify as central. In most such activities, school districts play a major role in determining the issues that the collaborative effort is to focus on. (2) Several of the superintendents participating in the Study, near the end of its initial three-year period, stated that in continuing collaborative activities they preferred to participate more fully in problem identification.

Since the Stanford Study was envisioned from the outset as research-based, there was strong reliance on and early attention to the collection of useful data. For two main reasons, concerted attention to gathering information about problems that beset the schools — by direct observations, by reviews of previous work, by surveys, by interviews and by other methods — was the initial step in all the projects: (1) Stanford's strength lies in its ability to conduct research: it selects professors on the basis of their strength in this dimension; it emphasizes research in its programs for doctoral students; its greatest credibility in the public-policy arena results from its professors speaking from a research base. (2) Data, if they reflect conditions that are considered important and interesting, provide an excellent stimulus for discussion, conjecture, and resultant collective action; they provide apparently objective raw material for focused interaction between University faculty and those in the schools.

In most instances, the form for data collection, as well as the specific data to be collected, was determined by participants from Stanford. The major exceptions were in the cases of the projects initiated by Professors Sanford Dornbusch, Hans Weiler and Michael Garet. In Dornbusch's case, the principals participated actively in the design of the survey instruments that were to be administered in connection with the Study. Dornbusch again:

> An example of their practical wisdom might be instructive. I had long used measures of cutting classes as part of an attempt to determine levels of student effort. What they pointed out to me was the class bias of focusing only on cutting. Students from higher social classes avoid classes without cutting; they use their extra-curicular activities to get excuses for absences from classes they don't want to attend. Now we ask about avoiding classes, as well as cutting.
>
> We have been meeting at least once every two weeks, and we go over aspects of our plans for learning about families and schools. The principals are not oriented to theoretical inquiries; they want practical knowledge that they can use. They constantly ask whether and how they can use our findings to help teachers and families to work better together. If we can't use the information, why discover it?
>
> So we scratch and claw our way to a mutually agreeable outcome, with no winners and no losers. Just imagine what it is like for me to hear objections to the wording of a questionnaire item from people who are sensitive to the nuances of interpretation among students, teachers and community members. They attack my outdated vision of the teenage world, and they do so with great precision. For example, when we asked working students to identify their job category from a list previously developed for a national study, they laughed at the naïvety of the researchers. The list no longer contained the usual jobs that youngsters hold. On behalf of researchers, I was a little embarrassed.

Still, Dornbusch told the principals at the outset that the survey instrument had to be of an appropriate length, that it had to be coherent and that such decisions, in the last analysis, could not be made collectively. His judgment would be final with respect to the data-gathering instrument. The principals understood the need for and accepted the ground rules.

Data Collection

Collection of data was no simple task. Professor Elliot Eisner wanted students' views of the high school. At what level are individual teenagers challenged intellectually, if at all? Is there much educational relationship between what happens in one class and what happens in the next? What general views do students have about the purposes and effectiveness of their own schooling? To gain insight into these issues, he decided to 'shadow' students in four high schools. The procedure was for graduate students to spend two weeks with each of more than twenty high-school students, observing them and talking with them during the course of the the full school day, and usually well into the afternoon and evening.

Such an activity can be disruptive. To ensure permission from the school administrators, to gain the confidence of the teachers, to select the students and persuade them to help, and to enlist the support of the parents all required an unusual degree of cooperation from many different participants. The process of building a cooperative relationship is itself delicate and time-consuming, apart from the sensitive issues that would arise during the shadowing, and afterwards. Yet permission and cooperation were secured. The shadowing proceeded as planned.

Another form for data collection was to ask students to respond to items on a questionnaire. In the project on personal responsibility, Professor John Krumboltz and his colleagues wanted students to rank different goals of schooling, like learning science or mathematics, or developing constructive health habits. Dornbusch wanted to find about the amount of homework youngsters were expected to do and the conditions at home for doing such school-related work. Different information was needed in other projects.

The potential for disruption was considerable if more than forty researchers (professors and graduate students) were to go into schools at individually arranged times and with uncoordinated schedules. Just as serious, the amount of time necessary solely to arrange for different groups from Stanford to collect data was awesome. In the end, it was decided that all the projects would collect data collectively. Each of the researchers would submit his or her needs for information to a central committee, a specially created Data Collection Committee. The researchers also would suggest how the information might be collected. The Committee, in turn, would screen the requests and try to design an efficient procedure that would minimize disruption in the schools and still collect the necessary data.

In a major and innovative effort, the Committee, led by Professor Robert Calfee, designed a 'shuttle', a probe that was to go into schools at a specified hour on a certain day to administer a package of questionnaires, suitably sampled, so that each of the projects would secure the information it needed but with minimum inconvenience to all.

Teachers and administrators welcomed this evidence of sensitivity on the part of the University professors to the potential for disruption that exists when researchers work in schools. Considering the huge amount of information collected during the shuttle runs those at the University, too, felt some satisfaction about the efficiency achieved, and in fact the procedure elicited favorable comment from researchers who work in schools elsewhere, along with expressed intentions to try the same procedure.

Once data were collected, the next step was to present the information in a coherent form to the teachers, principals and superintendents. Most commonly, the occasion for this process was a meeting between a researcher and a group of teachers and administrators particularly knowledgeable about the subject. The key question posed was whether or not the data presented a picture of a condition recognizable to those from the schools, the issue of verisimilitude. Did the data correspond with teachers' and administrators' picture of what was going on? For example, Professor Michael Garet, in studying course-taking patterns, found that students in roughly the top 20 per cent of ability seemed to be enrolled in similar courses; the titles, at least, were the same. Similarly, those in the bottom 20 per cent in terms of academic ability displayed consistency in their course patterns. On the other hand, no such pattern or consistency could be observed in the transcripts of those in the middle 60 per cent. In effect, their programs seemed like a random walk through the curriculum.

Survey instruments and transcripts analyses are one thing. The impressions of teachers and school administrators are another. Was the view from the school anything like the view that emerged from the data collected and summarized by University faculty and students?

The Study Takes Shape

The pattern of data collection and analysis described here was not envisioned at the outset of the Study in the form it finally took. The Study was massive and, in many ways, unprecedented. More than twenty professors, double that number of graduate research assistants at some stages, teachers and school administrators in twelve high schools operating in six school districts, and, at various times, well over 8,000 high school students (and some of their parents) were all involved at a significant level during the course of the almost-four-year period. It was far from clear at the outset either what themes would emerge as the Study progressed, or the precise form of interaction between those in the schools and those at the University.

After about a year, however, the sequence of collaborative activities described here began to take clearer shape and provided a pattern for the scores of people who would be involved in data collection and analysis. After problem identification and data analysis, four questions were posed in meetings involving University and school-district participants:

1   Does the information that has been collected correspond with the experience of teachers and school administrators in the district? Do the data ring true?

2   Does the situation highlighted by the data reflect a condition that should be changed? That is, do the data illuminate problems that those in the schools wish to address?

3   How do district faculty and administrators account for the findings? What circumstances have led to the current condition?

4   What lines of action, if any, seem appropriate in light of the findings?

In this progression of questions and in the subsequent deliberations about them, major initiative passed from those at the University to those in the schools. Professors and graduate students are good at collecting data; teachers and school administrators usually are not. Teachers and school administrators are practiced and effective at generating responses to meet daily problems in the schools; professors, as a rule, are not.

The data stimulated extended discussions about priorities, both educational and in the conduct of the Study. Was it educationally unsound that there seemed to be relatively little consistency in course-taking patterns for most of the students? Was it disturbing that children from single-parent families seemed to do relatively poorly in school, and that this factor seemed an even more important predictor than social class concerning how much a given student's achievement might differ from expectations? Was the issue important enough to command attention from teachers and school administrators in view of all the other demands on their time?

Another feature of most of the discussions between University-based people and those from the schools was an attempt to learn about the origins of the difficulty or condition that had been identified as problematic. To design a plan to ameliorate difficulties, it was important to understand how the problem had arisen. Was the apparent randomness of course choice the result of a point of view about electives, or a need to respond to the individualized interests of children, or an attempt to provide teachers with the courses they most wanted to teach, or something else?

For example, at a meeting of principals and a few superintendents convened to comment on Garet's work in progress, and specifically to consider the apparent randomness of course taking and course offerings in schools, the Garet group assumed that administrators and teachers simply were not paying enough attention to scheduling issues. They thought their observations and suggestions would be welcome and helpful. Instead they found that those in the schools were extremely attentive to scheduling, but that the University group were simply unfamiliar with the enormity of the constraints. For example, the administrators noted that they cannot receive State reimbursement for hiring teachers unless they identify thirty new students. But they do not know how many students will appear until shortly before school begins. In many fields, the turnover and enrollment fluctuations are sizeable — according to some principals, 'unpredictable within fifty kids'. Moreover they are limited in their ability to hire part-time teachers. Therefore, if there are thirty more youngsters in a class, they may need one more algebra section and two more English classes, but they can hire only one full-time teacher.

Additionally, the union contract limits the ability of those in the schools to adjust class size upward, and also the ability to move teachers from school to school. It also limits how many courses a single teacher can teach. One teacher cannot teach five different courses entailing five different preparations. The requirement is sensible, but it does add significantly to difficulties that might be seen initially as 'scheduling problems'.

As a further example, principals must keep certain courses for unusual reasons. A teacher of a vocational course about foods went on sabbatical for a year, and the school decided to drop the course for that period since it was not crucial in the curriculum or in demand among students. But late in the summer, the principal realized that if this particular course were dropped, the school would fall below the minimum number of courses needed to qualify for federal funds that support vocational training. So the course had to be reinserted in the program.

Thus, it was only in the time-consuming but absolutely necessary interactions between those at the university and those in the schools, everyone learned, that the research itself could lead to accurate and useful results. Analysis and prescription are misdirected if based on inaccurate or incomplete information, of course, but it is not always obvious to someone unfamiliar with the details of any practice where the errors or the gaps are to be found.

One of the most imporant outcomes of these discussions involving researchers and those from the schools was substantiation of the initial assumption in the Study that many of the problems that are found in schools are not the result of inattentiveness or lack of commitment to their resolution. Quite the contrary. Virtually every teacher and school administrator who participated in analysis of the data was eager to improve the quality of education for children. Rather, the problems reflect complex phenomena, for which ready solutions are not always at hand. The Study provided new opportunities for thinking about some of these troublesome issues, and, in the process of collective inquiry, led to fresh perspectives. Sometimes these perspectives, in turn, facilitated the development of new courses of action to make things better.

Chapter 4

Putting the Picture Together

The major goal of the Study of Stanford and the Schools was to draw Stanford University and several of the neighboring school districts into a new set of collaborative activities that had a central and clear purpose: to develop a picture of schooling in the high schools that participated in the Study, and to begin to formulate plans to attack the problems that were identified as troubling.

Rich and abundant information about the schools, as well as insightful observations, flowed from each of the many, separate research projects; there was intensive data collection and analysis across a broad front. To the authors of this volume fell the task of sketching the general picture, and we present the results in the next seven chapters — organizing our impressions around six broad themes. In a concluding chapter, we comment on the collaboration itself, and its future.

The separate parts of the picture came together slowly for us. We wanted, if we could, to distill some general impressions about life in schools from the research undertaken in the projects, but the task was elusive. Many features of the world we wanted to understand were revealed indistinctly at first, partly because of the piecemeal nature of the data collection that was an inevitable result of many research projects functioning simultaneously, partly because we did not try to force coordination on the projects at a conceptual level (telling ourselves that tight, central control would discourage potential participants), and partly because of the enormous amount of information that was collected.

A massive 'shuttle' surveyed thousands of students in one week on topics that ranged from their academic priorities to their part-time employment. Analysis of the resulting information took months and surfaced intermittently. Some researchers shadowed students, observed their classes, listened to the students' comments about their work in school, then spoke to the teacher of the courses and solicited their impressions of teaching and the student responses. Other researchers submitted questionnaires to parents of students in the same school. Several of the professors spoke to the school principal, or the district superintendent, and recorded their statements. In many cases, the principal, or superintendent or teacher were collecting data for the Study themselves and transmitting the information to the Stanford-based researchers. The sheer volume of information collected made it difficult to discern significant themes that might be viewed as transcending or adequately reflecting the work of scores of people.

Of course, each of the different projects had its own coherence, and each one resulted in its own set of observations: on testing, on the curriculum as experienced by the students, on technology in the schools, on change in the curriculum in recent decades, on students' course choices, on the influence of family structure and dynamics on the achievement of students, on teaching about America's interrelationships with other countries in the world, on the schools' role in teaching youngsters about personal responsibility, and on the changing balance of control in establishing a Californian state education policy. All of this work resulted in special reports, and much of it was published in professional journals, in newsletters and as book chapters.

Gradually, however, all this information — and particularly the distinctive voices and images that had been heard and seen — began to merge for us into a complex, multi-faceted and interconnected picture of today's schools. We must reiterate that the perspectives presented in the remaining chapters reflect the salient viewpoints that emerged for the three of us during the course of the Study as we followed progress in the separate projects and participated in the analysis of the data. We make no claim that the views we came to hold, and that we try to describe here, characterize education in other school districts, nor do we claim even that others who were examining or who might examine the same research activities would come away with identical impressions about the schools we did study. These are our own pictures, formed by our own experiences in and biases about the world of education — but shaped significantly by what we found out about schools, children, teachers, parents and school administrators as a result of the studies undertaken by our many colleagues at Stanford and in the schools.

We gathered considerable understanding from the several surveys done in connection with the Study. But, for us, a scribbled comment on a questionnaire sometimes shed light on a teacher's isolation or frustration; an observation of a middle-track student in class, gazing out the window at the trail of an ascending airplane, was often more eloquent than a statistical summary. We found ourselves influenced significantly by these glimpses of particular children and teachers as we tried to glean meaning from the masses of data.

Nevertheless, though we do not attempt to summarize here the various research projects undertaken in the course of the Study, we believe that the impressions that follow are grounded firmly in the observations and conscientious analysis of our colleagues in the schools and at Stanford who did all the work. We try to make the basis for that conviction clear throughout. We believe also that our observations are fully consistent with the results of our colleagues' analyses, as reported in their publications and private comments. The selections, however, and particularly the emphases, are ours alone.

In identifying and elaborating upon the various themes that are presented in the following pages, it was necessary, as Professor Dornbusch had told us, to penetrate a different world. The sensation was sometimes dizzying, and a discussion of the results of our Study would be too disembodied without at least a modest attempt to portray the pace and pressure in today's high schools.

First, the school world is a faster, more intensified form of daily life than we are accustomed to, or that we think is typical. Days pass in a blur of ringing bells, relentless schedules and pent-up, explosive, teenage insecurities, hostilities and sexuality. It's a world where 'crisis' is commonplace: the books don't arrive for a certain course, a student is arrested, a teacher goes to the hospital and no substitute can be found for a course, an angry parent wants an explanation for a low grade, marijuana is found in a teenager's locker.

It's a world where the difference between an A- and a B + can make a teacher's telephone ring, precipitate a student's emotional crisis or launch a parent's vendetta against a school — and it's a decision of the sort that must be made almost every week, by most teachers, for about 150 different students. Also, in a typical high school, 2,000 courses selections are made each term. Yet there is little time to think, little time to do anything but react.

Professor Elliot Eisner pointed out that a high school teacher has more direct contact with youngsters in formal classes in one month than a professor has in three, and his estimate may be low. A university professor, often with the assistance of an assistant, assesses perhaps two papers and exams from each of his students a year. A high school teacher may grade daily homework assignments, alone; that teacher teaches five course periods a day, forty-five minutes each, with up to forty new faces in each class.

While the world of the high school is in many ways a troubled and troubling one, then, full of conflict, spiritlessness, disappointment and confusion, it is also a world in which teachers and students display optimism, humor, compassion, energy, ability, commitment, dedication and, often, profound self-understanding. Granted, we were predisposed to find the latter set of characteristics. Nevertheless, we found it in every one of the schools in which we worked. It was gratifying, and it was unmistakable.

Before we could begin work in earnest, then, we first had to adjust. Our adjustment to the exigencies of school life must now become the readers' adjustment, too. Without it, the often-contradictory, high-pressure and troubled picture of high schools becomes too easy to criticize, too difficult to understand and impossible to change.

Chapter 5

Contradictions in the Schools

Consistency is not always a virtue, and most people even have learned to value a certain level of ambiguity in their organizational as well as in their personal lives. Nevertheless, we began to take notice as the Study progressed of the number of contradictions between the stated aims of the school and the practices within them, and the dissonance we detected emerged as one of our main 'themes'.

Such contradictions, some fundamental and some apparently trivial, were identified by teachers and researchers in each of the projects. We were not shocked to find evidence of this sort, given what we had heard about schools from writers on educational reform in recent years[1]— and given, also, the size of high schools and their mix of students, the differing expectations for them by different audiences and their funding levels in view of their many goals. Considering also their intense and busy pace, with little opportunity or encouragement for teachers and administrators to sort out even the most glaring inconsistencies, it probably is no longer news that there is a noticeable gap between the schools' aspirations and public declarations, on the one hand, and school realities on the other. What did emerge forcefully from the Study, however, and what disturbed us in its intensity, was the acute awareness and distress with which some teachers, students, researchers and some of the parents sensed the contradictions.

In their studies of testing in the schools, Professors Robert Calfee and Edward Haertel found that most teachers have broad, idealistic, over-arching goals for their students. Typically, they found, an American history teacher cares deeply that students grasp a sense of the origins of Western culture and understand some of the broad patterns of its subsequent development. Virtually every English teacher wants the students to capture something of the greatness and variety of our literary heritage, and also wants them to be stimulated to read more, and enjoy it. And yet, when it comes to evaluating what teenagers are learning in school, the teachers often test students only on acquisition of relatively unimportant and isolated facts that require little more mental activity than rote recall. Teachers seldom try to assess deeper understanding or appreciation, or how well the students are able to relate their most recent reading to other things they have learned.

The reasons seem blunt and straightforward, both to Stanford observers and to the teachers themselves; test questions demanding simple recall of discrete bits of information can be administered by means of the kinds of multiple-choice exams that later can be graded electronically, saving sizeable amounts of teacher time. It is much harder to develop multiple-choice questions that test understanding of the broad concepts on which teachers say they place greater value. Larger class sizes (which are a particularly acute problem in California) and an increased number of administrative tasks are absorbing more of the teachers' time than was the case ten years ago, leaving them less opportunity to develop and grade examinations that assess deeper comprehension.

The contradiction between the rhetoric of educational goals and the reality of classroom practice in testing, however, creates a further contradiction in the students' minds. Although teachers, parents and political leaders tell students about complexity and the value of education in understanding the subtle relationships and distinctions by which we guide our lives and shape our understanding of the world, what schools in fact demonstrate to them is a misleadingly simple and understated view of the effort required to lead a thoughtful life. According to Calfee:

> These kids think 'education' is a collection of trivial facts. And they don't expect to learn these facts unless they have been spoon-fed them in preparation for a test . . . Students don't see that education is going beyond the 'right' answer.

Professor Elliot Eisner noted in the course of the Study that we *say* learning is an important goal in and of itself, but we place an astonishing emphasis on a whole range of external motivators to raise student performance. According to Eisner:

> One of our schools had a five-and-a-half page, single-spaced list of points given for offices held, exhibitions in which students had participated, differential points for coming in first, second, third and fourth

in those exhibitions, points differentiated for type of school
service . . . Are our children being subtly hooked on point acquisition?

In other words intellectually questionable entertainment was used to support
education. In another instance, he noted that a school used distracting and
noisy pinball machines to raise money. Professor Dornbusch's work revealed
that external rewards undercut the process of education: where parents use
money, cars and privileges to change the behavior of their students, children's
grades suffer. He is not sure of the reasons, but postulates that these external
motivators send messages to children that 'education' isn't important, grades
and prizes are — and internal motivation is thereby reduced.

In the course of the Study, we found that students are extremely sensitive
to contradictions in what we say and what we do. Repeatedly, on the questionnaires, they showed great awareness of the lack of adequate school facilities,
or funding, or the other, subtle messages that undercut our protestations that
we value education. One student, describing the problems at his school, also
expressed his understanding of the reason: 'Our school's broke!'

When high school seniors are about to graduate, they are rewarded for
their perseverance with Senior Skip Day – in other words, they are allowed a
day off from school. When a student cuts classes, and is caught, he or she is
punished by suspension. We offer the same treatment as punishment and
reward. What do students make of that contradiction?

We say that education is important — and yet the school facilities we send
them to, since the Proposition 13 cutbacks in California, are, as one young researcher describes, 'scuzzy'. What do students think of the ostensible importance of education when they are sent to facilities that are dirtier than their
homes, overcrowded and badly in need of repair?

We say human values are important, and that education is important, but
witness this sensitive recollection by an English teacher and active researcher in
the Study, Charlotte Krepismann, who watched a British film on slavery in a
colleague's history class:

> I found these scenes very painful. . . . Yet even here, students' internal
> sense of time had them filling the room with the sound of zippered
> backpacks before the bell rang — and while human beings with black
> skins were being beaten like dogs. Where have our students learned to
> zip up their sensibilities with the sound of the bell? They know that
> outside this classroom lies another one where being tardy is frowned
> upon and nowhere in the new district regulations on tardiness does

one find an excuse for a student who wants to think about or even discuss material he has just absorbed. Perhaps, if he's lucky, he can find someone at lunch who may want to discuss the disturbing film or lecture. Meanwhile — on to the next show . . .

One researcher records a gym teacher's call to his students, as he asked them to line up according to height: 'Know your place in the world!' he shouted. Ironically he was blithely exhorting these nervous adolescents to master the knowledge that has eluded great philosophers for centuries. (We might say the same of the schools' perceived role to help students 'be at harmony' with themselves.)

Not all the contradictions are as transparent as these, and many reflect the larger contradictions of the society as a whole. Professors John Krumboltz and Martin Ford conducted extensive surveys among parents, teachers and students. They found universal agreement about the role of the school: it is supposed to do everything. In their study, parents, students and teachers were queried about the goals they felt represented the most important outcomes of schooling. All parties agreed that a moral sense, positive attitudes about education and the ability to function effectively and considerately with other people were the most important areas to be mastered. At the same time, all respondents concurred that schools also bear most of the responsibility for developing students' mathematical and verbal abilities. These opinions were consistent across boundaries of gender, academic track (for students), race, socio-economic standing or schools.

There was little awareness that schools may not be able to assume the role of primary agent in reaching all the worthwhile goals of a productive life. When asked specifically about the function of schools, all three groups (parents, students and teachers) insisted that schools should have a major, or at least moderate, role in developing a student's sense of right and wrong as well as in encouraging sound health practices, and the development of skills in working with other people.

Traditionally, engendering such desirable characteristics in the young have been seen more as the responsibility of the family or religious institutions than a self-conscious goal of formal education.

Martin Ford concluded: 'I think our results say clearly to the school: "You should feel comfortable addressing issues of moral behavior, particularly public behavior"'. But the question then arises about the appropriate emphasis on such topics in the formal and public curriculum of the school, as contrasted with the responsibility of the school as a somewhat self-contained community to exemplify, in the way people associate with one another, the morality and behavior that are judged to be desirable. The following comment by a parent, reflecting a type of experience that probably is not uncommon, underscores the contradiction between the rhetoric and the reality:

> Last week I found out my son (14 years old) was having a problem in his PE class. I asked my son if he had mentioned to his teacher that this was the third time his PE clothes were stolen and my son's answer really upsad (*sic*) me. He said, 'Mom, I told my other teacher, but they don't care. My PE class is very big, about sixty students, and every three weeks we get a new teacher'.

Krumboltz also found that even the goal of learning to get along with one's family is considered a responsibility of the school by 81 per cent of the students, 96 per cent of the parents and 91 per cent of the teachers. Ford concludes:

> People seem to recognize the school as a general social force. So it does have some responsibility for a wide variety of outcomes. Whether it tries or not, it will have effects on some of these social and moral areas. So it cannot avoid responsibility in these areas.

It is perhaps ironic that the school is seen as such an omnibus institution at a time of general financial cutbacks in education, a California crisis in the building and maintenance of school facilities (as described in the next section) and a relatively low level of teacher salaries, at least during the three-plus years of the Study. Schools are supposed to do it all — and yet we do not give it the funding of an organization expected to be so influential: we expect it to teach the 'basics', provide vocational guidance, prepare students for college, train them to be good citizens, reinforce their sense of moral obligation and, at the very least, keep them off the streets and out of trouble. In a number of cases that we found surprising, teachers are expected to serve as parents, filling-in for family members in helping young people with all sorts of problems, including their social relationships with their friends, and even difficulties with the law.

Because these problems are serious, and because there is often no one else to provide assistance, and simply as a consequence of a compassionate response by one human being to another, teachers and school administrators try to help. Many of them do not see it as a matter of choice when students turn to them with their troubles. Yet teachers and administrators are given neither the pay nor the status — nor do they usually have the training — for a role so demanding, open-ended and important.

While the expectation is broadly shared that schools can and should do almost everything that is deemed worthwhile in society, when it comes down to cases, there is often conflict; consensus seems to disappear when issues arise that are controversial in the broader society. The contradiction had become commonplace, but nevertheless striking: although people wish schools to encourage sound health practices, when schools actually assume the responsibility of teaching about AIDS-prevention, for example, some members of the public voice pronounced agitation and even anger. We wish schools to teach citizenship and morals, and yet the recent legal battles over sex education and religion in textbooks demonstrate that when schools actually attempt to teach values, many people react sharply and negatively. Witness also the persistent conflicts about the teaching of evolution and about school prayer. Witness the repeated attempts by some groups to ban certain books from school libraries and excise material seen as 'amoral' from others.

One student's comment on the Krumboltz questionnaire directed anger toward schools and their demands, university researchers and their queries, and may more generally reflect students' fatigue with trying to fulfill society's expectations, with its possibly idealized vision of itself:

> If these self-righteous morons would stop trying to teach us to be 'open and caring' or to be 'excited by learning', the schools would be much better. I can handle my moral and spiritual growth. I don't need school prayer on the right and psycobabble (*sic*) from the left. I need to know how right (*sic*) and count, not how to 'maintain loving relationships' or to 'be happy with myself as a person'. And as for 'getting along with different ethnic and cultural and racial groups', how many blacks live in the upper class areas of these babblers? Maybe two. And they're probably maids.

The questionnaires that were distributed to parents had a section for 'comments', so respondents had an opportunity to amplify their views about the topics on which they had been queried. Some parents took the opportunity to note that instruction about morals and 'personal' guidance should be the responsibility of 'parents' or 'the home' — but relatively few said it was 'my' responsibility, or, 'I am teaching my own child morals'. We believe it possible that this language reveals a more fundamental contradiction: the public may expect a long-standing and respected institution, like the 'home', to have some power beyond what we invest in it. That is, some parents may believe that 'home' carries certain desirable attributes, like stability, love and support — regardless of the realities of today's family life. Consequently, they believe certain things *should* happen at home, even if they seem actually to happen there less than might once have been the case. By default, the public sees responsibility drifting toward the school and supports that trend because the responsibility must reside somewhere.

Krumboltz's survey showed that parents, teachers and students want 18-year-olds to have achieved a sense of harmony with themselves, a moral sense of what's right and what's wrong, and a sense of self-esteem. But we seem not yet to have sorted out the key question of whether or not it is reasonable to expect the schools to play a key role in all this, at least formally.

'Church' has little power to teach values that the public wants instilled in children unless young people attend. Do the many parents who cited the church's role in developing a moral sense actually attend a church? We don't know, but figures on church attendance suggest that very large numbers do not. Are parents who cite the 'home' as the most-desired shaper of children's sense of right and wrong actually making the effort to teach the values they espouse to their children — or do they rely solely on the lessons that will be effortlessly and wordlessly absorbed, lessons that may correspond poorly to the values that the parents would choose to state as representing their most deeply held beliefs? The figures on teenage pregnancy suggest that children are not learning about the sexual behavior most valued by the community at home — parental protestations and aspirations to the contrary.

Sadly and ironically, the written remarks of parents sometimes revealed the spelling and grammatical errors they were criticizing the schools for not eliminating in the written work of their children. ('I don't have a college degree', wrote one, 'but I can still tell they aren't learning as much. I know there is budget problems, which I personnelly think are ridiculos, but it's just an excuse'. (*sic*)) Perhaps there is a deeper issue here that centers on the images that people have come to acquire about education and schools. Perhaps this demand of the schools is another example of the persistence of an unrealistic vision of what these institutions can do, at least at present. Perhaps parents so desperately wish for their children what they themselves have been unable to achieve and cannot provide that they turn to other people and institutions to fill the void.

If so, the phenomenon is not new, and there is awareness of the fact that the schools historically have demonstrated their ability to help bring children to levels of achievement that far surpassed those of their parents. But there is at least a possibility that the gap between increasingly explicit parental expectations and the ability of secondary schools to meet those expectations is of a different order today than was the case a generation ago — both because the secondary schools serve a much broader base of the student population now and because many homes seem less able to provide the educational experiences they once did.

But it is possible that only in this generation has the educational dream acquired such as non-acadmeic (and perhaps non-material) overtones: the quest for a sense of self-esteem, harmony, moral judgment. It is unclear to us which outcomes parents and students *really* wish schools to strive for – and how much the teachers are willing to assume this burden.

Unfortunately, we did not ask what functions are to be foregone if almost everything is a worthy goal. If the answer from the community is 'Nothing', as we suspect it is, then the public must be willing to support the schools much more than it is now doing for these institutions to have a reasonable chance of success.

Note

1   For example, THEODORE R. SIZER'S *Horace's Compromise* (Houghton Miflin, 1984), JOHN I. GOODLAD'S *A Place Called School* (McGraw-Hill, 1984), SARA LAWRENCE LIGHTFOOT'S *The Good High Sshool* (Basic Books, 1983), and ERNEST L. BOYER'S *High School* (Harper and Row, 1983).

Chapter 6

Who's Running Things?

Education is a mammoth endeavor in California: the State enrolls over 4.1 million students, employs more than 200,000 teachers and disburses about $17 billion in annual expenditures. California is divided into 1,029 school districts, which range in size from a half-million students (in the Los Angeles Unified School District) to seven students (in the Fall Creek School District in Siskiyou County). Forty per cent of these districts enroll fewer than 500 students.

And yet, local superintendent Paul Sakamoto asserts in a document prepared as part of the Study, that the State enacts legislation for all schools as if every one of them were part of the 'Golden State Unified School District'. Professor Michael Kirst's work, in conjunction with Sakamoto, chronicled how local school districts have gradually been losing control to State-level authorities — a development that has been largely unplanned by legislators, apparently unwanted by the public and probably educationally harmful for children in many school districts.

'We're looking at a thirty-year trend that's been caused by many events', said Kirst. 'The bottom-line conclusion is an imbalance: and excess of State control and a striking lack of local authority',

Sakamoto recalled the days when physical education was the only curricular requirement in the State Education Code. In the 1985 legislative session, however, the state passed 183 bills and concurrent resolutions, ranging from one prohibiting stun guns on campus (AB2191) to adding one semester of economics as a high school graduation requirement (SB1213).

But we are not entirely sure what, in fact, 'State control' means, and what the consequences are. Constitutionally, of course, the State has the authority to pass legislation to govern educational practices. However, it bears only remote and indirect responsibility for the results of those laws. When a child fails the newly mandated economics course, the legislator in Sacramento is not the one to receive the parent's angry complaint, even if the reason is lack of funds provided by State-level authorities to hire a competent teacher of economics. Rather it is the local teacher and school administrator who bear the brunt of the blame. It is they who must tailor a response by making the necessary adjustments in the school's program or in the course of the child's education.

Who controls the schools, then? Laws are one thing. Providing individualized educational services to specific children is something else. Who, rightly, bears responsibility for the quality of education — or, more precisely, who bears responsibility for which elements of the educational enterprise and how is the public to understand who is accountable for what? The confusion, conflict and complexity resulting from the lack of clarity make it unclear where responsibility lies for fashioning remedies. This fact emerged during the course of the Study as another revealing lens through which we now view the problems of schools.

After nearly four years of research, the verdict, as expressed by Kirst, still has the power to disconcert: 'There are no villains here. There is no "Master Plan" '. Each person acts on the parts of the educational 'system' closest at hand, those that are easiest for that person to manipulate. A teacher changes an assignment to better suit a particular child; a superintendent hires the best teacher he can find to fill a vacancy; a principal tries sensitively to understand the basis for a parent's distress; the State schools' superintendent tries to mobilize the political support necessary to provide adequate educational funding; a legislator writes a law; the Governor juggles political priorities. Everyone is well-motivated. Many are able. Most work very hard.

But, in fact, schools seem to be drifting, with no one at the helm. There is no consistent direction; the mandate seems to change from year to year, certainly from election to election. At one time, it is basic education. At another, science and math. Later, it is cultural literacy. Yet, none of these concepts is understood well, not even the teaching of science and mathematics: Do we teach children the science necessary to enter college, or do we teach more about the uses of science by the non-specialist citizen in daily life? The two are not the same, yet both demands are made on the schools at one and the same time.

Only the general institutional inertia existing in any long-established body maintains a steady sense of purpose, coherence and stability — and sometimes that picture is illusory. School staff remains largely the same from year to year; their lesson plans remain intact; children continue to attend classes; even standardized achievement tests seem to show little annual change. But these signs of apparent stability may mask slow and deep changes that, over time, can be profound. For example, if one looks at a twenty-year period (rather than the more common two- or three-year perspective), there is considerable evidence that the dramatic expansion of the educational mission of American high schools, to serve all the children and not solely those headed for college, is working.

Increasing State direction of schools, without concomitant direct responsibility, leaves many school administrators feeling helpless, concluding that they are no longer in charge of the school and its curriculum, but are spending increasing amounts of their time adjusting to pressures from far away — with little sense sometimes that the adjustments they must make are in the best interests of the communities they serve.

During the course of the Study, researchers were confronted with many stories of State-level, and even large-district-level, intervention in instructional matters, which was often so hastily developed and general in its scope as to lack meaning at the school-site. Frequently, State or district mandates dictate what legislators, school-board members or even staff in the State Education Department want to happen, but without consideration of how a school can manage the required changes. The problem is predictable in view of the diversity in California's schools. Still, the political pressures to act are immense, even when the results of that action are not always clear. And just about all that a politician or school-board member can do is pass a law or implement a regulation.

A particularly revealing anecdote emerged at one point in the Study: to increase instructional time for students, one region established the requirement for an additional twenty-eight minutes in the school day. But no one had provided the District with funds to increase the pay for teachers, now required to add half-an-hour's teaching time to their work day. Teacher pay was but one unaddressed issue; another was the need for buses to be rescheduled. But the school had a unique reaction to avoid other problems: administrators added two minutes to the breaks *between* each course period. No one had meant to subvert the policy-maker's intention; nor had the policy-maker meant to disrupt the school's operation.

Perhaps a more typical example of the problems produced when new educational practices are imposed from a distance surfaced during an interview with a district superintendent who had just received news of legislation requiring one term of economics in high school. 'Where the hell do you find someone to teach economics?' he asked in exasperation. 'It's a complex subject!' Not only was it necessary to recruit one new teacher for the curriculum (in competition with all the other State's high schools, which would be looking for people with the same qualifications), somewhere they had to budget for an additional salary. Suddenly, the course scheduling mechanism, already a nightmare of variables, had one more element to be factored in.

In this case, there was no particular quarrel with the educational value of instituting a required economics course. Rather the superintendent felt that those who had established the requirement had little knowledge of the complexities of schooling and had made inadequate provision for all the consequences of the immediate implementation of the new law. Furthermore, the new course was injected into the program with little knowledge of how it related to other aspects of the educational program, or of what it might displace — thus adding to the tendency to build curriculum by accretion, a matter to which we return in a later chapter.

Some Stanford researchers had their own quarrels with these State mandates — quarrels usually grounded in the content, rather than the confusion and lack of adequate preparation for the State's new and more assertive role.

As Professor Calfee noted: 'Although the State is trying increasingly to determine what is taught, their mandates are not rooted in what the curriculum is, nor in any idea of what the classroom is'. These broad-gauged State initiatives tend to cluster around remedies that are inexpensive and relatively easy to administer. In reality, most of the effective responses to some of the real problems that have been identified are expensive, long-term and difficult.

Calfee cites the California Assessment Program (CAP) testing as an example. Public uproar over whether the average scores increase or decline a few points means as little as the test scores themselves (in view of what the tests actually measure), but can deflect needed education dollars into unprofitable avenues. He concluded, 'It's yielding poorer education, not better; everything in this disconnected system is moving toward worsening education'. Overstated, perhaps, in a moment of frustration. But it has been clear in many settings that laws and regulations are blunt instruments for effecting sensible changes in the administration of personal services, whether in education, or in medicine, or in social work. The problem is one of getting the right balance between the public will as conveyed through elected officials, on the one hand, and professional judgment as expressed by teachers and school administrators, on the other. Schools cannot operate effectively without public support, and not solely financial support. Nor can they be run sensibly without ample scope for professional judgment about how individual children are best served.[1]

Not surprisingly in view of the difficulties, attention within the school sometimes focuses on ways to deflect outside mandates that teachers and principals consider counter-productive — especially requirements from the State, that occasionally seem particularly detached from the realities of school life. In the words of Professor Michael Garet:

> The administrators we observed at the four schools displayed remark-
> able dexterity in dealing with the numerous uncertainties involved in
> planning the curriculum. Over the course of the curriculum planning
> cycle, large and small events outside the school — new district guide-
> lines, new district administrators, increased state requirements, disap-
> pearing federal funds, new mainstreaming policies, new college entry
> requirements and revised teacher contracts...compete for the
> attention of administrators, placing new demands on the curriculum
> or re-establishing old ones. The administrators we observed accept
> this 'noise' as the occupational norm and have learned to formulate a
> curriculum plan and carry it out while devoting a combination of real
> and symbolic attention to external and internal demands.

We do not mean to portray teachers and school administrators as guileful. They aren't. Rather, they are trying to make workable, local adaptations to mandates that usually do not reflect accurately the precise needs of their schools or districts. Teachers might question, and some have, even how much the local-district authorities, the boards of education, are in touch with their needs.

Today's practitioners often do not feel they are in control of their own destinies. Mandates from the State rain down upon the county and districts; the principals sometimes feel oppressed by directives from the distict office; the teachers are known to express resentment of principals. According to Super-intendent Sakamoto:

> County officers feel they could do without the State, and that money
> should be allocated directly to them. We at the district feel we could
> do without the county, and schools probably feel that they could do
> without my office . . .
> There is a lack of trust — of the legislature in us, the local school
> district — that we will spend the money as it was meant to be spent.

As a result, Sakamoto reports that his district must wait until July or August to see how much money it will get for the school year that begins in September. Thus, the district has to negotiate pay raises, for example, without knowing whether it will have the money to meet the obligations. The feeling at all levels of power that we studied is one of increasing helplessness, and distrust of any outside intervention. Said Stanford researcher John Agnew, who is also a former high school teacher: 'More trust might be built if people stopped messing with them, if they really had autonomy: the principal over the school, the teacher over the classroom, the parents to hire and fire the principal'. But no such autonomy exists, as least we did not see it.

The School Board

Kirst's work chronicled how local school boards, the primary agent of local control, have had their most important prerogatives assumed by the State. For almost all of their history, school boards played a central role in developing programs and dealing with the basic curriculum issues: the subjects to be taught and the time devoted to each one. Now these matters are decided in Sacramento.

School boards, however, are left to deal with issues like dress codes, smoking and open campuses. School board membership is still an extraordinarily demanding and largely unappreciated task, with accompanying monumental public pressure. But many of the aspects of the work considered most rewarding are no longer present. Sakamoto concludes: 'I think there are no really positive reasons to run for election on a school board nowadays'.

Apparently, others agree with his verdict: in Sakamoto's Mountain View/Los Altos Unified School District, only one non-incumbent ran for office in a recent school board election; it was a LaRouche candidate, who had never been to a single board meeting. 'That election cost us $32,000', notes Sakamoto, 'even though there was no serious opposition to the incumbents.'

The Principals

Much legislative reform depends on the high-school principal for faithful translation into a program that matches political intent. But the principal's ability to make changes, within his or her own school, is limited, even illusory.

Dornbusch noted that 'no free-floating power' exists in the school. No one speaks with clear and unchallenged authority, and principals must lobby constantly for support from their teachers. Researchers in the Study repeatedly offered examples of principals who 'laid themselves on the line' by supporting the collaborative project with Stanford, sometimes without prior support from the teachers who would be required to participate in interviews, fill out questionnaires, use class time for surveys, or, in many cases, collaborate as researchers. When a teacher decides actively to oppose a principal, the opportunity for that teacher to create mischief, or worse, is immense. If a teacher fails to give students the correct directions, for example, the results of a survey can be severely compromised. Moreover, they can oppose a principal later, when a serious educational issue needs action. For principals, commitment to change can be unusually stressful, and lonely.

The Teachers

A teacher's control even over the classroom is constrained. Because teachers give grades, they are subjected to intense pressure from some parents, and even to threats from students. Examples of the pressures that teachers must face were noted often during teacher interviews and, occasionally, on the margins of questionnaires. Scrawled one:

> What crazy parents do is manifested in what the kids do. You have crazy parents, you get crazy kids. The mother said: 'I can't accept anything below a B'. I'm going to call her tomorrow to tell her he hasn't turned anything in yet. Then the mother calls and does her trip.

Liora Bresler summarized similar findings from the Eisner project on curriculum:

> ... virtually all of the teachers expressed feelings of being unsupported, alone and often helpless in their efforts to promote student learning. Overall, these teachers perceived guidance from their departments, schools and districts as largely non-existent.

Much in the same vein, researcher David Flinders, reporting on a group's interviews with teachers, noted:

> Teachers typically spoke of district guidelines which 'nobody follows', textbook approval procedures as 'not rigid' and competency testing as

'something I go beyond'. Nevertheless, teachers expressed mixed feelings regarding their classroom autonomy. On the one hand, they felt their freedom to be necessary to their work and one of the few positive aspects of public school teaching. On the other hand, almost every teacher we spoke with attributed their autonomy not to professional status, but rather to indifference and a general lack of support. Shortages of materials, facilities and money were repeatedly mentioned by teachers as the primary constraints on curricular decision-making. One teacher, for example, said that he was forced to use two different textbooks in teaching the same course due to inadequate supplies. Other teachers pointed to dwindling human resources by noting the elimination of district curriculum specialists and the continual loss of their best faculty to better paying jobs outside education. An art teacher commented: 'Nobody's in charge at the district. Lots of the art program is chopped to shreds now. It's a real disaster story'. This same teacher also reported that he receives only $4.64 per student per year for supplies and equipment, money which, in his words, 'doesn't go far'. Another teacher whose department has been cut back from a faculty of five to one-and-a-half said he feels he and his students have been 'put on the back porch'.

The Physical Plant

Kirst's report notes that California school enrollment is growing at the rate of 100,000 pupils per year, and is predicted to continue at that rate for at least the next seven years. He reports that facilities have already reached a crisis point in southern California, where schools are not air-conditioned and are severely overcrowded. Some are more than 50 per cent over capacity. The Los Angeles School District has started year-round schooling to avoid building new facilities. From an interview:

'They can't build buildings or fix their own facilities. They're wholly dependent on the State for construction funds. But the State funds are exhausted,' said Kirst. In several areas where school buildings are needed, 'a local construction bond issue could pass overwhelmingly — but local citizens can't vote on them. The State says we can't tax ourselves anymore. A parcel tax is insufficient for this type of construction.'

But the situation is not always so critical, nor so regionalized. A stroll through any local school — even in a 'good' school district, like Menlo-Atherton High School — revealed years of neglect. Menlo-Atherton, the site of one of the Study's annual conferences, struck one observer this way:

> The building had a shabby look. Paint was peeling off the walls and ceiling. The lockers, also in bad need of repainting, had been scrubbed so often the metal showed through. Menlo-Atherton, like so many other California high schools, had continued only 'crisis maintenance' since the passage of Proposition 13 (a California tax reduction bill) had slashed their resources.

Another observer, cited by researcher Liora Bresler, describes a different high school this way:

> The physical environment reflects and reinforces a 'non-caring' attitude: dirt, neglect, graffiti, garbage, clocks that are hours late or early (impossible to tell which . . .), shabby faculty rooms with hard, uncomfortable furniture and depressing surroundings. Noise pollution is often manifested in the din of outside traffic and classroom telephones which interrupt lessons continuously, reflecting a lack of respect for the learning process. (Compare this, for example, with the silent atmosphere of our churches and concert halls.) The physical environment as reflecting classroom values is a recurrent theme in our research.

Researchers weren't the only ones to cite public neglect. Sakamoto also notes the decline in his Study paper, 'The Golden State Unified School District':

> Many of our schools are 20 and 30 years old, and are in need of major maintenance. Waste paper containers are used to catch drops of water from leaky roofs. Classrooms are often too cold to teach in because of faulty furnaces. Cracks in the asphalt of the parking lot grow wider and deeper with each year of neglect. Student desks have been repaired so often they will not stand more repair.

For Kirst, his examination of eroding local control and increasing State control during the several years of the Study led to a sharp shift in his general orientation toward the distribution of power between local and State authorities. Kirst was one of the early, influential figures in American education who advocated a shift of power toward high levels of government. He exerted this influence during his years in Washington as the staff director for the US Senate Subcommittee on Manpower, Employment and Poverty; as the director of Program Planning and Evaluation for the Bureau of Elementary and Secondary Education; and as associate director of the National Advisory Council on Education of Disadvantaged Children. He is also a former President of the California State Board of Education. Says Kirst of the shift in his fundamental outlook:

> My government career has stressed intervention from higher levels to the local school districts. In 1965, in Washington, I argued that local schools were paying too little attention to civil rights, equity for the handicapped and the dropout situation. In the late-70s and early-80s, I advocated a massive intervention to raise academic standards. But several years ago, I began to look at the total State role and became concerned. It's not that I'm an idealogue for local control — I'm not.
>
> Individually, the interventions have been very good. Most of them, frankly, I support. But their collective effect is dangerous. It's a case of too much of a good thing.

Too much of a good thing? Perhaps. But the questions remain: Who will take responsibility for the various levels of accountability at the school-level, at the state level, or at the national level? Most importantly, who, ultimately, will insure high-quality education at the level of the student?

These questions, however, presuppose clearly drawn lines of authority. More than that, they imply that various bodies have the power to perform their assigned tasks, whether such power means decision-making opportunities for local school boards, or providing funds to schools and school districts from a higher governmental level. We found it difficult throughout the Study to find unambiguous lines of authority, or identify many areas in which the exercise of power was not seriously questioned.

Assigning blame is not the aim of this chapter; but noting what happens as a consequence of educational policy, intended or not, is. The end result, in this case, seems clear to us: neglect. Neglect that, in some cases, verges on the incomprehensible in view of the unquestioned importance of schools, their role in healthy development of the State and Nation, and the rate at which they must grow in the years immediately ahead. Yet, we came away from the initial years of the Study unsure of where pressures must be applied to make the necessary changes.

Note

1   For an expanded discussion of 'education by remote control,' see ARTHUR E. WISE'S *Legislated Learning* (University of California Press, 1979) and J. MYRON ATKIN'S 'Government in the classroom'. *Daedalus*, Summer, 1980.

Chapter 7

Stability and Fragmentation

We were reminded as we talked with participants in the Study and began to examine the data that were being collected that life is highly fragmented in today's high schools: knowledge is divided into uniformly sized fifty-minute segments. Furthermore, there is little attempt to integrate the total school experience into a coherent perspective on our heritage, or on our understanding of the physical universe, or on the political and social events that confront people today. This observation should sound familiar to those who have followed recent reports on high schools, like those cited on page 00. It should sound familiar also to almost anyone, anytime, who has attended an American high school.

During the course of the discussions between Stanford-based researchers and the teachers, however, we noted that this ubiquitous and prominent atomization of life and learning in schools seemed also to serve as a significant force for insuring stability. Dividing the day into period-sized pieces makes the school more manageable, and may give it the day-to-day stability to transcend enormous potential perturbations, like staff and student turnover, as well as the daily, unanticipated emergencies. As Professor Michael Garet noted:

> The . . . curriculum decision-making process we observed acts as a mechanism for managing uncertainty and constraints. Furthermore, the process has a kind of inertia that can keep a school moving when a shake-up or illness temporarily leaves the school without leadership. In short, the process serves to buffer the school's curriculum (as well as teachers and students) against the ebb and flow of the internal and external environments.

For example, when information is fragmented into discrete fifty-minute segments, it allows a teacher who suddenly quits, or falls ill, to be replaced by another teacher who is able to teach the same subject. Teachers become somewhat interchangable, and the entire structure of the school is less subject to severe disruption.

Every organization needs a degree of stability — high schools, perhaps, more than most — not only for reasons Garet presents, but because of the much more peripheral consideration that virtually everyone spends many years there and likes to view the institution afterward as relatively unchanging, as a fondly (or not-so-fondly) remembered and fixed feature of one's youth. The extent to which such non-academic expectations are met, in fact, is probably directly related to the amount of support, financial and otherwise, that the public is prepared to provide to the schools.

The question that then arises is: Can we strike a suitable balance between fragmentation and stability? Professor Elliot Eisner believes that fragmentation has gone too far: 'Students study American history and American literature as if these two worlds had nothing to do with each other'. He describes the educational consequence of this separation:

> A curriculum that isolates science from its cultural and historical context diminishes its relevance for those who are unlikely to regard the sciences as appropriate for their professional or vocational aims. And that means most students.

However, most knowledgeable observers can understand and support the advantages of dividing the high-school day into manageable parts, not only because of the necessary subject-matter specialization required of the teacher (that must be accompanied by at least some degree of compartmentalization), but also from the point of view of appropriate educational quality and range for the student. A student's Spanish class potentially may relate to his social studies or history class and, given the demographic swing in the country, even to his current events class. But such subject-overlap will not bring any greater understanding to the hours of drill-work needed to master foreign grammar, nor enliven the long list of vocabulary words that must be memorized. In some ways, greater integration of themes across subject fields may even serve to distract a student from the effort needed for certain necessarily isolated, but important, learning.

Nevertheless, there certainly is a line that, when crossed, leads to education that becomes too disjointed, too incoherent to the learner, too en-cyclopedic in intent and effect. At that point, schools lose their clear educational goals and risk descending to intellectual and emotional chaos.

They become something other than primarily educational institutions, perhaps facilities for substitute child-care. Is there a danger that that point has been reached? We wanted to find out. How do students view their schools, educationally?

Researcher Carol Leth Stone's observation of the college-bound 'overachieving' student she shadowed seems to bear out Eisner's conclusion:

> I [asked him] whether he saw any connections among the various classes he is taking. He obviously thought that a very strange question. To him, each subject is in a separate compartment, and it had not occurred to him that they could be perceived or taught in another way . . . He studies now so that he can study in the future.

Witness also this reaction from Charlotte Krepismann, a veteran, highly praised high school English teacher who, in Eisner's project, spent three full school-days with a high school student as part of the 'shadowing' effort. It was one of the few occasions she had ever had to take a more detached, and possibly more objective, view of high school education in a career that had spanned decades:

> I can only compare it to spending several days on Broadway visiting one theatrical production after another, each one with a different director. Of course, in my analogy, there is no need to tie the various parts together nor to tell the jaded audience what it was supposed to carry home after all the stimulation. Yet theater often presents some way of relating to what we all think of as the 'real world', and, even if we subject ourselves to an overdose of the stage world, one part of our mind glimpses that real world through the shadowy one. In school, where we are ostensibly preparing our students for that step into reality, I saw no attempt to make connections between each fifty-minute segment of the students' journey through their day, and, only in certain obvious cases, was there an attempt to point out the importance of each day's work to some larger scheme of events.

This disjointed and unconnected approach to education for high-school students is not a simple matter to rectify. High schools have been organized by subject specialization for at least a century (and with good reasons). A teacher lectures, a bell rings, and students proceed to the next teacher, the next lecture. The growing number of students now participating in the educational system, despite their differing needs and differing destinations, have embedded these habits even more deeply into the educational system, creating an organizational tangle easier to maintain than to simplify and change.

Scheduling Pressures

What are some additional manifestations and consequences of fragmentation? As noted, in today's high schools, everyone is subjected to the schedule. Students' routinely adjust their educational plans to fit its requirements, rushing to meet a September deadline of course assignments: If two courses that they need meet at the same hour, there is no choice but to alter their educational goal, at least in a small way, but often in a big one. Teachers and administrators struggle to keep abreast of faculty and course changes, which are in continual flux due to staff and student turnover. Garet observed:

> The number of decisions that have to be made at a high school each year is staggering. In a school with 2,000 students, and with each student taking six courses, that means 12,000 course placement decisions have to be made every term. And on the average, each student makes two additional course changes later on during the year.
>
> In addition, most schools maintain standardized achievement test scores as well as proficiency exam results. Overall, the number of individual pieces of information involved in curriculum decision-making at the school level is staggering.

With co-researchers Brian DeLany and John Agnew, he traced the decisions about courses that were taken for a year at each of four participating high schools in different districts.

In DeLany's words, they found that:

> the process of coping often leaves the administration with very little elbow room in planning curriculum. The administration is caught up with trying to cope with limited time and resources, because when September comes, the thing starts. You can't say, 'Wait kids! If you come back a week from now, the books will be in!'

As a result of the frantic pressures in scheduling, staffing and the acquisition of materials, school officials are hampered even in their ability to monitor the courses that individual students take. Of the four high schools Garet and his colleagues studied, two no longer had counselors even to fulfill nominally the role of supervising student course-taking, thanks again to Proposition 13 cutbacks. Schools that had counselors weren't much better off. Typically, a counselor was assigned to oversee 600 students, which means that virtually all attention must go to those who take the time to solicit help, or to those whose behavior demands that the school take action to avoid disruption.

Garet noted the 'the school's *intention* is to check and make sure everyone is taking the appropriate courses', but time and staff are simply not available for much follow-through. 'We *expect* students will be appropriately matched with courses. But if we want careful, appropriate placement, then schools need the organizational support to do it. They need additional staff and better information management'.

DeLany witnessed a school's course-change procedure, and compared it to a line-up at a busy store, where customers take numbers and wait for the attention of a sales clerk. 'Two hundred students came, until administrators ran out of numbers and had to turn them away', he said. 'The students were handled very perfunctorily. They were all processed in an hour.'

No large school can ignore or avoid the mammoth difficulties entailed in scheduling. Every student must have somewhere to be, during each of the six class periods, at the beginning of each term. To maintain its own legitimacy as an accountable, public institution, a school must attend urgently and immediately to that requirement, it seems to us, even though it may absorb extraordinary amounts of the school's time, and detract from the school's responsibility to educate.

'Meeting this deadline becomes as big a problem as what the schedule might mean to student and teachers', said Garet. We think he has argued persuasively that much of the difficulty of educational reform lies in the bureaucratic barrier between the courses teachers, parents and students want, and the courses they eventually take.

Garet also found that turnover played a significant role in increasing levels of distress and confusion in schools. More than ever, we are a mobile society: between May and October of the year he studied, turnover ranged from 14 per cent in one school to 27 per cent in another.

In other words, he said, 'In a school with 2,000 students, every year there are about 400 kids who are scheduled to attend, but suddenly aren't there, and maybe 450 others who suddenly show up in September to replace them. It is exhausting for the school.'

The number of students replacing the departing students is not always balanced: one school had a net decline of 137 students in a little over a year.

Moreover, even slight variations in student population can upset the schedule: the number of students taking an advanced course may drop below an acceptable minimum, or the number of low-achieving students may unexpectedly rise, making the addition of another remedial English course imperative. The school does not always have the staff to manage the last-minute fluctuations.

Such sudden crises require reshuffling the carefully balanced courses. Students who had artfully juggled their schedules so that they could take trigonometry and chemistry may suddenly find both offered only during the same course period.

However, Garet found that some schools have learned to be inventive with the high number of students who neglect course request forms: they are used as 'fill-in' for already scheduled but under-enrolled courses. Of course, these assignments may not be helpful for the students used as filler.

The Myth of Tracking

Though high schools track students into either general, college preparatory or vocational courses, Garet found that these terms often fail to describe a student's actual schedule:

> You'd expect kids with different goals to be taking different courses appropriate to their goals — but that's less the case than you would expect. In fact, there's a good deal of overlap, even apparent randomness, in what kids take and what course sequence they follow.

He and his co-researchers spoke at the 1986 Conference of the American Educational Research Association with what may have been the most intriguing lecture title at the meeting: 'Down with Tracking — Whatever It Is!' As the title suggests, Garet and his colleagues emerged from the Study with severe doubts about how effective tracking is — or whether it exists at all. 'Kids aren't divided up into neat boxes', he said. Part of the explanation may be that the school simply doesn't have the resources to follow-up to make sure the student takes courses that are consistent, one with the others, in reaching a desired educational goal.

Neither do students seem to know whether or not they are following an educational track to a clearly identified destination. For example, Garet found that substantially fewer high-school seniors reported plans to attend a major university immediately after graduation than did ninth-graders. This may be a result of modified aspirations during a period of three years, but it also may be attributable to lack of clarity among the students; many ninth-graders are unsure how to characterize their program of study. When asked to describe their high-school program as college preparatory, general, vocational or remedial, more than a third of the ninth-graders responded, 'Don't know'. In contrast, almost none of the seniors responded, 'Don't know'.

Garet believes that the process of articulating an educational goal is not so much one of careful decision as it is one of default — and gradually discerning the handwriting on the wall. As he put it:

> The questionnaire data we have analyzed . . . leads us to believe that, for many students, course choices are only loosely connected to educational plans, and the educational plans are somewhat unstable.
>
> Our data on student course-taking patterns cast doubt on the common view that high school students are neatly separated into two tracks — college and non-college. Instead, our data indicate that student course-taking patterns are quite diverse.

For example, at one school, students followed more than sixty routes through the science curriculum. The most common one was followed by only 16 per cent of the students. Garet also described an apparent randomness in the students' course-taking patterns:

> You get that feeling as soon as students try to explain why they're in the classes. Some explained it was because they couldn't get into another course, either because the course was full, or because it conflicted with something else. The only other course open might be physics or French III, and that may not be in the universe of courses the kid would take. You could have a course open, but irrelevant.

'Tracking', however, is still very much alive in the quality of education a child receives once they have chosen a course, however accidentally the decision was made. David Flinders, studying the curriculum of one English teacher who conducted three college-preparatory courses and two classes for non-college bound seniors, found that quality and emphasis differed sharply between the courses considered 'college-bound' and 'non-college bound':

> In one semester, the college-bound class will study twenty-three authors, read thirty-eight short stories, poems or essays. They will

read and be tested on seven complete novels, and write ten short essays and one research paper. They will see or listen to twenty-six films or tapes. In addition to this, they will receive twenty-eight reading quizzes, eight unit tests, and eight unit vocabulary tests on 220 words.

In contrast, the non-college bound are given two days a week for unassigned reading, may pick a novel to read (usually gothic romances or thrillers), with one vocabulary lesson and one vocabulary quiz each week. At the end of the semester, they are tested on a total of fifty vocabulary words.

Revisiting Stability and Change

On some days, in some settings, to some observers, high schools seem to be bastions of resiliency. They absorb, adapt and often eventually reject changes, always maintaining an identifiable continuity, a very persistent stability. As we have said, in fact, one effect of reading the many documents chronicling class-room observations, teaching styles and especially the shadowing reports of students during the course of the Study is to note how little schools have changed, how familiar much of it all seems — at least when we put ourselves in the students' place. Multi-colored, punk mohawks may have replaced yesterday's crew-cuts, but the curriculum and psychology of students (though less with teachers) seemed curiously arrested in time. This sense of familiarity was evoked among many readers of the Study's reports and among those who did the actual observations, from twenty-five-year-old graduate students, to sixty-year-old professors.

And yet on other days and through other lenses, schools seem to have changed tremendously — as Professor Decker Walker found when he decided to study changes in high schools programs during the last twenty-five years. He had not expected to find great change in high schools when he started; he, too, had a sense that schools had not changed much in the last quarter-century. The principals of the collaborating high schools, by the way, agreed with his perception.

But using the twenty-five year time-frame, Walker found 'remarkable changes in scheduling, number of courses, content of courses and courses never taught before — such as Bilingual and Advanced Placement'. Further, 'There have been a tremendous number and variety of changes', he concluded, 'and most of them are still around, at least in name'.

When Southeast Asian immigrants began to flood into certain Bay Area districts, the schools responded with classes on English as a second language. When the students of the 1960s demanded 'relevant' courses, new topics such as black and minority studies were inaugurated. Language laboratories were built (and later abandoned). Team teaching was tried. 'New Math' was introduced. Girls' sports programs were initiated in response to the public's desire to foster equity between the sexes. There was much more.

For example, in the period from 1970 to 1975 alone, in the schools that were studied, thirteen elective courses were added in seven departments, some in response to student demand (psychology, film, TV production), others in response to teacher request (minority history, American studies, aerospace); a career center was established; a school/community service program was expanded; and a broadening of course offerings included student-led seminars in math, classes in graphic arts, metabolic physiology, individualized research projects for advanced science students, independent study opportunities in core subjects, English electives geared for different levels of proficiency (including Poetry of Rock, Sports and the American Scene, The Formative Years), physical education electives (Bowling, Figure Control, Skin Diving), new social studies courses (such as Minority Cultures, Youth, Law and Society), mini-courses in Braille, computer, the metric system, photography, and sign language. These are only some of the changes. Teachers and high school administrators were steadily accepting changes that kept the curriculum and resultant scheduling at a constant boil. Why?

Robert Palazzi, Principal of Aragon High School, offered a list of reasons: new principals often reorganize the school when they arrive. Major funding discontinuities, such as Proposition 13 or School Improvement Program grants, contract or expand the supply of dollars for new or existing programs. Staff turnover causes change, since a course may be added or cancelled when a new teacher comes or an experienced one leaves. Moreover, schools try to be responsive to the ever-changing demands of parents and students.

As for students, Palazzi commented: 'We went through the "sixties kid". Then there was the return to "the fifties kid". And now it seems we may be seeing a resurgence of the questioning "sixties students" in the freshman class.' And curriculum will change accordingly.

But Walker also found that the principals were correct when they stated that little had changed, for though a set of courses around the periphery were in constant flux, the academic core courses remained the same. These are the courses most students take, for most of the day. Consequently, despite a course selection that may have dazzled their parents, most students do not feel they really have tremendous range in course selection: most of them, most of the time, take the familiar courses they are supposed to take, or that are required.

Garet found that students do not think of these choices as options on which they must take action: 'These decisions just happen. It's what you do when you're a high school student. You fill out forms, you get courses. It's like thinking about what to have for breakfast. To call it a "decision" would be inflating it.'.

Why the same core courses? The heart of the matter may lie in a comment by Dr. Robert Madgic, Director of Curriculum and Instruction for the Mountain View/Los Altos Union High School District. At a Stanford colloquium, he said that the basics have not changed in several decades 'because we don't know what's a better way for students to learn'. He pointed out that schools have no easy 'bottom line' to measure success. A business can turn to profits and losses. But how does a school measure its 'profits' and 'losses' against changing socio-economic conditions, shifting demographics, fluctuating funding, transitory staff, sudden influxes of non-English-speaking immigrants and, finally, the changing native ability of the student population that must be served?

Chapter 8

Testing in the Schools: The Tail Wags the Dog

Professors Robert Calfee and Edward Haertel focused in the Study on the relationships between testing and teaching. In surveying more than 250 teachers, they found that at least 10 to 15 per cent of the class time available for instruction is devoted to testing. In a school year of 180 days, spread over ten months, that works out to almost a month and a half of school time each year for each student in which the sole activity is responding to questions on examinations. In some classes, the figure jumps to well over 20 per cent. In smaller-scale terms, that means that about half a class period each week in every class is spent on testing, or more than one test a day for every student.

Calfee and Haertel's numbers swell if the time spent on discussion when tests or quizzes are graded and returned to the students is included. Moreover, their figure does not reflect the amount of teacher time spent devising and grading tests and quizzes. Nor does it reflect the amount of time students spend outside of school preparing for them. When 1,888 surveyed students were asked to rank the importance of their test scores from 1 (not important) to 5 (very important), 83 per cent indicated 4 or 5. Eighty-four per cent of the parents gave the same ranking, as did 62 per cent of the teachers. More potently, perhaps, half the students claimed that they usually or always consider their test scores when deciding on a career and whether to consider applying to college.

Underscoring the role of tests, we found in the Study that many courses seem to be organized around the tests. Typically, courses are divided into 'units', which, in turn, are usually structured around the textbook. These units vary in length from three days to three weeks. In virtually every case, the unit ends in an examination; for longer units, there are tests on sub-portions. According to Haertel: '. . . many students perceive teaching and testing as alternating in a dreary cycle: a unit is taught, then tested; the next unit is taught, then tested, and so on. Material that has appeared on a test can be forgotten.'

In a sense, then, the whole curriculum could be said to be geared toward the test. In some cases, like the Advanced Placement preparatory classes, this process is even more explicit: the entire course is designed around an examination that will be taken at the end of the year.

Then there is an entirely different realm of testing, one with which the public has become quite familiar during the last two decades: the 'standardized' tests that so often make headlines, including, for example, the Scholastic Aptitude Tests (SATs), the California Achievement Program (CAP) and minimum competency tests.

Gradually, we began to note that today's schools are driven by assessment to a degree that surprised us. Having worked in educational institutions all our lives, we are not unfamiliar with the fact that tests are given (though the amount of testing was striking). As a result of the Study, however, we began to believe that the tests may be used as ends in themselves, rather than as one means to assist in intellectual development of the individual student, or even as a reasonable method of public accountability.

Tests — usually standardized tests, but even including tests that teachers devise to monitor and judge week-to-week progress — are omnipresent and constantly on everyone's minds. Teachers, administrators and students are preoccupied with scores and grades — often against their better judgment, and often against the values that parents, students and they themselves say that the schools should reflect, as we have seen in an earlier chapter.

Standardized Tests versus Teacher Testing

Calfee makes a sharp distinction between what he calls 'the two faces of testing': the standardized tests; and the more subtle, complex, and personalized tools, teachers use in their own classroom to measure student achievement and that function powerfully in assigning grades. Haertel observes:

> As you enter the world of classrooms, there is a sort of *Alice in Wonderland* or *Through the Looking Glass* reversal — the meaning of the word 'test' changes, and standardized tests suddenly seem remote and insignificant.

Standardized tests may matter little in the [daily] lives of students and teachers, but other kinds of tests do matter. The classroom tests that teachers ponder and students study for, that bring closure to each unit of instruction, that determine students' grades and help teachers judge their own effectiveness, become critically important.

In the schools participating in the Study, only 1 per cent of the high school students even knew what the California Assessment Program was.

While publicity about average scores on standardized tests receives national attention, the other tests, which may, in fact, be truer witnesses to a child's education, are not usually considered by the public at large, however seriously they are taken behind school walls and by the parents of individual children. Outside, they are taken for granted, assumed as one of the routines of school life. According to Haertel:

> When children enter school, testing is part of what they find, like taking attendance or collecting the milk money. By the time they reach high school, all the quizzes, review sheets, unit tests, finals, labs, themes, reports and presentations are just a part of the routine. My first impression when I began to analyze the results of our student and teacher questionnaires was that classroom testing was almost completely non-controversial. When asked how much testing there should be in their classes, three-quarters of the students chose the response 'about as much as there is now'. When asked to compare test-taking with other school activities, their modal response was that 'taking tests is no better or worse than other class work.' For most students, testing is okay — not great, but okay.

By contrast, a third of the surveyed teachers thought that standardized achievement tests were not worth the time and effort to administer. Another third were unsure.

Calfee and Haertel, as the Study was in progress, noted the gradual encroachment of State-developed tests on the school curriculum. State-wide and nationally developed tests were introduced slowly, cautiously and experimentally soon after World War II for specific and well-understood purposes. Gradually, however, the uses to which the tests were put began to shift.

The Scholastic Aptitude Test (the SAT), for example, was devised to predict an individual student's success in college. Now, however, they have to be used for a number of additional purposes, none of which they were ever intended to serve, and all of which are disturbing to the test-makers themselves: to compare schools, districts or states; as a measure of accountability to the public; or to evaluate specific curriculum programs. Since not all students take SATs, only those with college aspirations, and not all states even require them for admission to publicly supported colleges, gross comparisons are meaningless. Still, they have become one common indicator by which the public regularly and simplistically measures the quality of education. Professor Michael Kirst has termed them 'the Dow-Jones Industrial Average of Education'.

At many points during the Study, researchers observed the effects of our national preoccupation with testing in the schools, and heard many rumors. CAP testing, in particular, provided a rich source of anecdotes and impressions. The tests, because they are used to compare schools and are a factor in State funding ('Cash for CAP'), are now highly competitive. More than one researcher returned from the school sites with stories about how schools elsewhere had boosted their scores by encouraging low-achieving students not to attend school on the day of the CAP examinations. In another unnamed school, according to another rumor, twelfth-graders took revenge on the school's administrators, who had cancelled the traditional Senior Skip Day, by deliberately scoring poorly on the tests.

For tests used to screen applicants for college admission and where student motivation to achieve is therefore higher, other means of distorting the ostensible aims of the tests were commonly noted. Special coaching courses now are routine for the SAT and the ACT (American College Testing), and almost certainly help to raise scores, despite occasional challenges to that belief from time-to-time by the test developers. Teachers complained of them, both because of a perceived element of unfairness (some students take the coaching courses and some do not) and because the tests are not intended to be responsive to cramming.

Calfee finds today's standardized tests deficient in content: 'At present, they test fairly trivial things in a fairly trivial way', Calfee said. For example, he cited one standardized writing test that gave students only two minutes to write a short essay on a passage of literature. 'It's a surface-level response', he said.

Nonetheless, these tests have encroached powerfully on the schools and on teaching. And at some point, they began to shape the curriculum — an effect that was not originally intended. Calfee noted that teacher-devised tests are beginning to resemble the format and style of standardized tests. For example, textbooks and teacher's manuals generally include questions teachers can use for assessment. Some may require the student to prepare an essay or write out the solution to a problem. But in many instances, the format has become multiple-choice or short-answer, requiring only a word or phrase from the student — just like a standardized achievement test.

Perhaps these were some of the examinations Professor Elliot Eisner had in mind when he wrote this condemnation of high school tests:

> They are too narrow, they neglect personal forms of achievement, they encourage educationally conservative practices and they direct our students' attention to the wrong goals. One of our most important tasks is to invent better ways to reveal to the public what they have a right to know, namely, how do we as professionals know how their children are doing as students? . . . The aim of curriculum and teaching is not simply to help students meet the demands of schooling, but to help them use what they learn in school to meet the demands of life outside school. . . . How can we move away from programs and methods and incentives that breed short-term compliance and short-term memory?

Tests and the Curriculum

Many teachers and school administrators agree that the emphasis on and uses of testing reflect an educational practice out of control. Points, credits and scores are pursued and accumulated as if they represented the core values of the high school. Because of their prominence, grades become the focus of many educational discussions.

One teacher complained: 'Some parents do not have realistic expectations. You have kids who get a B + and break down and cry. I say, "That's a good grade," but they say their parents want them to get an A.' The comment was not an isolated one.

Students, too, drive the curriculum toward the tests. First, the sheer number of them plays a role in determining the techniques teachers employ to assess what has been learned. There are many ways to evaluate levels of student understanding: listen to a child develop and present an argument; read a student's exposition of a theme; watch a performance; engage the student in conversation about the topic about which understanding is to be gauged.

However, with today's classroom sizes of forty students, many teachers gravitate toward multiple-choice exam — even for courses like algebra or literature, which are best tested by revealing the process of thought, rather than solely the result. The apparent reason? Such tests can be graded by student assistants or electronically. 'Students "buy off" on this', said Calfee. 'After all, studying for an essay exam is a lot of work.'

Teachers are aware of this process. As Calfee's researchers found in extensive interviews, teachers have articulate, idealistic educational goals for their students, but amidst the pressures they face, they proceed to test in predictable ways, often modeling their approaches on the externally developed examinations they see most often, the standardized achievement test. Or they simply use the tests included in textbooks.

Students exert a pressure beyond their number, and it reflects their view of the purposes of schooling. Almost all classroom observers noted that students frequently ask, 'What's going to be on the test?' or interrupt lectures to ask, 'Is this important? Is this going to be on the test?' In many cases, review for a test meant the teacher coaching the youngsters about the specific pieces of information to be included on the exam, along with the appropriate answer. This technique reduced the students' role to straightforward memorization. One researcher even recalled students asking a teacher: 'Are you going to cheat and include something on the test you haven't told us about?' Said Haertel:

> The problem is that students use test scores to figure out what they should be learning. Basically, kids think that if it's not on the test, it's not really important. They use tests to figure out what teachers consider worth learning. Students need to understand that pencil-and-paper tests judge one important, but limited, set of schooling outcomes. If they understand that teachers use other kinds of measures — homework, written work, class discussion — it might broaden their concepts about what outcomes are expected from them.

Furthermore, students often seem driven by a need to be evaluated, to know their standing. Such a desire is surely an understandable and prevalent human trait. But the students' method, almost exclusively, is to calculate grade-point averages (GPAs), double-check the points on a graded test and add up the credits toward graduation. Unable to judge themselves by their own standards, they form their perceptions of themselves based on techniques derived from authorities and peers. In school, that means, largely, examinations. Predictably, they prefer 'hard' black-and-white measures, clear and general indicators, to the more subtle, subjective technique of, say, a narrative assessment — partly because narrative evaluations are more difficult to compare, one with another. And they pressure schools and teachers to give them such assessments. For teachers, too, the same choice is tempting: it is easier to justify the grade based on multiple-choice quizzes than to try to define the qualitative differences between A − and B + essays.

The influences are circular. Students and parents affect teachers. But the patterns of assessment that prevail in the school, in turn, shape the students' views and expectations.

Said researcher, Liora Bresler, during a presentation on the Eisner panel findings:

> Our case studies suggest that students see the school experience primarily as a means to secure a diploma — a ticket for success for the future. School is not expected to be enjoyed and appreciated for its intrinsic benefits, but is regarded as a system which delivers grades. The high-achieving students aim at A's, and low-achieving students hope to pass; both groups have clear goals and know exactly what to do in order to reach them. Both groups accept narrow expectations regarding their own learning.
>
> Today's students do not comprehend the purpose of education, as these aims are usually discussed by educators and policy makers, nor do they usually acquire a taste for learning as it is understood by such people. Lining up points and credits that carry them toward college or a diploma is the universal surrogate.

Researcher Rebecca Hawthorne, who shadowed a student named 'John', noted in her report:

> Inside John's notebook is a photocopied handout entitled 'Require-ments of Graduation'. Conscientious reference to this sheet before selecting classes each semester has structured John's school existence. Discussing the *quality* of accumulated points and units makes John un-comfortable.

Calfee echoes the same sentiments: 'These kids think "education" is a collec-tion of trivial facts. Students don't see education as going beyond the "right" answer. Life is not a multiple-choice exam.'

The simplicity and ends-orientation of their thinking was reflected in students' remarks on Professor John Krumboltz's questionnaire: 'Don't put foreign language and "fine arts" in the same category. Speaking Spanish can get you a job. Comprehending the subtleties of Picasso's Blue Period cannot.' (Then the comment ended, less definitely, with: 'Do you realize that many of the students filling out your survey have little or no idea of what they're doing, and that subsequently a good percentage of the data you are collecting is completely meaningless?')

Calfee recalled meeting a teenage student who said US history was her favorite subject. For Calfee, it was an opportunity informally to assess her all-round educational progress. 'She didn't know much about the Revolutionary War. She didn't know much about Vietnam . . . . And when I asked her about Joseph McCarthy, she said, "Oh yeah, he was a famous general". That's a better reflection of what kids are learning'.

Anecdotal evidence — certainly. But it presents a vivid picture of what statistics are already showing us: our future voters do not know their cultural history, where they came from or where the future is likely to take us. In Hans Weiler's survey of 950 students, only half could name US Allies in World War II. Only a fifth knew the purpose of NATO, in a multiple choice response that showed almost a random spread. Many do not even know the rudiments of reading and arithmetic. Again, we have plenty of anecdotal evidence. Two parent comments from our questionnaires:

> In my work, I come in contact with high school graduates who cannot write a simple sentence. Many can barely write their own name . . . I've listened to many high school students who cannot read above a third grade level. . . .

> I was in a shop buying a Gloria Vanderbilt designer shirt. I commented that I didn't understand why the name 'Vanderbilt' gave her any more design ability than anyone else. The comment was lost on the clerk, about 24, and a [nearby] high school-aged shopper. They didn't know that the name Vanderbilt meant anything other than tight jeans. My stepson knows nothing of 'the robber barons' — why? But he plopped a can opener on the sink board and said it was Art, 'if he said so' . . . He passes tests beautifully, but I do not believe he can write a coherent paragraph.

Alternative Forms of Assessment

We believe there is a strong relationship between the kind of testing prevalent in schools today and the shallow conception of education held by most youngsters, many parents and even a small number of teachers. While the two phenomena are intertwined, and there is a bit of a chicken-and-egg problem, we think that useful effort could and should be applied to assist teachers with evaluation efforts that reflect their deeply held educational goals.

Many researchers and teachers longed for a set of achievement measures that would show the overall knowledge level of students, and how they can make use of it in thinking about their world. Calfee is one of them.

> For example, a school district could take a class hour in every school on one day to ask all eleventh-graders to write an essay on Chernobyl [which had occurred a week before]. That would give a good, all-round flavor of what kids are learning in a number of subjects: social studies, English, world events, thinking skills, attitudes.

In Calfee's hypothetical test, how students treat the various aspects of the subject — political, chemical, social, biological, historical — would show how well they can synthesize knowledge and form an intelligent reaction to events occurring all around them.

Nor are researchers and teachers alone in wishing for some way to evaluate what happens to a student over four years of high school that probes more significant outcomes of schooling than the mastery of discrete and isolated bits of information, many of which not only have little relationship one to another, but to little else besides. Collaborating Superintendent Paul Sakamoto had proposed a research agenda, which he characterized as the 'Fruit Cocktail Project'. The fruit cocktail metaphor stemmed from Sakamoto's early experience as a fruit cocktail inspector at a canning factory during the summer months:

> In those days, there would be a quality control person to check how many cubes of pears and peaches, how many grapes were included in each can. He would make sure each can had two halves of a maraschino cherry, and he would check the sugar content of the syrup. But at the end, the inspector would check the total product: the color, the texture and the taste.
>
> That's where we're missing out in educational research. I'd like to see someone look at the student after four years of high school in

total and ask: 'What impact has school had on this person generally? What is the overall final result?' This might include in-depth interviews — perhaps with parents, too.

We keep *so* many records on each student — like achievement tests, interest inventories, career plans and so on. But no one has pulled together all these records to look at the whole student. We're treating students in segments — in secondary education, we're still counting sections of the fruit.

We are still counting, and grading, sections of the fruit. But for more than one university researcher, and for us, these observations evoked pangs of conscience and self-doubt, in addition to feelings of distress about educational quality. After all, universities like Stanford, with their grade-oriented entrance requirements, certainly played a role in enshrining the results of tests and GPAs in ongoing educational policy. There is not much doubt that universities like our own contribute to the problems we were studying. In fact, many of the practices we found ourselves lamenting in the high schools not only stem, in part, from university requirements, but the practices themselves are found, at least to some degree, at our own University.

Chapter 9

Classroom Heroism — and Isolation

For a high school teacher, the first day of work after graduation from a university and certification by the State is often only slightly distinguishable from the last day before retirement: the same sea of teenage faces, often asking the same questions, requiring the same knowledge, dreaming many of the same dreams. Teachers receive little help from anyone else in the educational system during the course of a career in developing new and potentially useful perspectives on the challenges that face them, and only slightly more assistance with new techniques. Much of what they learn depends entirely on their own efforts and experience, which are confined, largely, to the boundaries of their classrooms and the lives they choose to lead outside their professional worlds. They learn, and derive encouragement and discouragement, from the students — but very little from other adults. The situation caused Professor Elliot Eisner to lament:

> Even ballet dancers who practice their art to perfection have mirrors to see for themselves how they are doing. Where are *our* [teachers'] mirrors? . . . Their mental health, the time they need for reflection, their skills as teachers are of the utmost importance. . . . No effort to improve schools can succeed unless our teachers are given the time, space and feedback — both critical and supportive — they must have to make collective educational aspirations a reality for students.

Contrast this need with the reality, outlined by Liora Bresler, one of Eisner's colleagues:

> Our findings support other classroom studies suggesting that teachers rarely receive collegial or administrative feedback, are somewhat isolated in their classrooms and often feel that no one supports or is

interested in their work. Consider the following examples: Mr. Dixon, a newly appointed English teacher, complains that nobody introduced him to the other faculty members, nobody showed him around, gave him the necessary keys, or explained to him on the first day where everything was. Few teachers greet him good morning, and after a year-and-a-half of teaching in the school, he was asked by another teacher if he was a substitute.

Neither do teachers often receive acknowledgement for doing better work, like improving tests and curriculum materials, or developing new ideas. Duties with students during weekends, evenings and summer vacations is sometimes required of teachers, but frequently not defined as work responsibilities in their contract. Some of the teachers complained that they rarely see their principals, and even more rarely obtain any useful information about their teaching from them. Virtually all teachers expressed feelings of being unsupported, alone and helpless in their efforts to improve their work with their students. In some cases, we found that the teacher's personal and professional involvement declined with experience. Moreover, many teachers commented on the fact that the public has a low esteem for their work, and that they are generally regarded as non-professionals. Still other teachers complained of lack of money for basic supplies and equipment, like up-to-date textbooks and curriculum materials.

Our case studies corroborated our interview findings in a related dimension: many teachers have been teaching at the same school for over twenty years, and there have been so few young, novice teachers brought in that some of the veteran teachers feel that teaching is a profession in decline.

Despite this gloomy picture, however, virtually every researcher involved in the Study who interviewed or observed teachers returned with anecdotes of the commitment, sincerity and idealism of the average, *not* the exceptional, high school teacher. Most, in fact, nearly all teachers involved in the Study had compassionate goals for the children, even when they could not fulfill them; even teachers characterized by others as 'early psychological retirees' expressed genuine affection for their students. Many followed the progress of their students after high school, through college or career. A very large number in the schools we worked with were the objects of pilgrimages by students after graduation.

Moreover, all reports came in with tales of classroom heroism: of dogged perseverance for excellence beyond what any society reasonably could ask.

Many teachers went above and beyond almost anyone's concept of even a well-developed sense of duty in counseling students, or spending extra time with them. No surprise, perhaps, since most adults can recall such teachers from the days of their own attendance at school. But it is always heartening to be reminded of such dedication, especially when the press is full of reports of educational incompetence, as it is periodically, and when teaching as a profession is so clearly underpaid, emotionally stressful, time-draining and socially derogated. Some of the more seasoned Study staff were moved to comment that the only reason they could identify to remain in the classroom, in view of the intense pressures and modest extrinsic rewards, is to fulfil an independent and powerful sense of mission, and feel the accompanying pride.

As Eisner expressed it:

> . . . We have inadvertently designed a system in which being good at what you do as a teacher is not *formally* rewarded (students provide rewards to teachers, but that is another matter), while being poor at what you do is seldom corrected or penalized. This, it seems to me, is a recipe for inertia.

However, he goes on to note: ' . . . schools are still populated by teachers who, despite these negative factors, perform with great artistry, indeed, brilliance, in their classrooms'.

An article that appeared in the Study's newsletter, *Update*, cited the following outstanding examples collected from Eisner's curriculum project alone:

● In Mr. Fasman's U.S. History classroom, bulletin boards teem with class photographs. But the photos are not the usual row of faces. Instead, they depict class enactments of American history: witnesses in military uniform testify about U.S. involvement in Vietnam; convicted Salem witches hang their heads; preachers rail against evil; and dour-faced judges preside over famous legal battles. For Fasman, all these trappings are a way of making history come alive for his students. 'What happened? That's *always* the question to ask yourself', he says to a class. And occasionally he puts students on the spot to create instant drama: 'What's going to happen as a result of the muckraker's journalism? Take it one step further. What's going to happen to Standard Oil? Pretend you're the public in 1903 and you're *mad! What are you going to do?*' Fasman is a popular teacher. In the opinion of one of his students, 'This is a mind-exciting class. It makes you think'.

● (American literature teacher) Mrs. Muffler's classes are highly engaging, and the students pay attention, according to Stanford researcher Sigrun Gudmundsdottir. 'She leads them almost step-by-step through each chapter, using quizzes and assignments to get them to read and pick up the topics that she wants identified. These assignments are vocabulary lists, study guides and reward-the-reader questions', says Gudmundsdottir. 'Moreover, she relentlessly makes sure they have no excuse for not reading. She explains, clarifies and carries extra books so everybody can read when she wants them to read in class . . . . There is nothing timid or hesitant about her teaching. She launches enthusiastically into every lesson as if this moment were the climax of her career, and she carries this dynamic feeling right through the lesson.

● Mr. Hill [is] a good science teacher who patiently tries to teach the school's 'hot potato' — a general science course that no teacher wants, with an unruly class that doesn't want to learn. Or Mr. Dow, a science teacher who engaged his class in a lively, unresolvable discussion about 'How do you define life?' (The group finally concluded that there is no real answer — fire comes closest to being definable as life without being life.)

● Mrs. Hainsworth [is] an English teacher whose unofficial goal for her course is, whenever possible, to get students to empathize with an adult perspective. During a period when students were reading *Ordinary People*, for example, Mrs. Hainsworth brought in a guest speaker. Mr. Cooper, a retired English teacher who used to teach in the school, talked about his son's drug problems. Mr. Cooper told a stunned class how his son, a high school football star, became a drop-out from life. The talk was emotional, the class responsive and it was another triumph for Mrs. Hainsworth's unofficial goal. The students saw the 'other side' — the perspective of concerned parents who often take the overprotective, nay-saying role in their own lives. And through it, of course, they gained an empathy they can bring to their own understanding of literature.

Most of the highly praised teachers we came across during the course of the Study had dramatic classroom flair. They lived life at an elevated emotional plane, making it obvious to all who met them that they cared deeply about what they did. We began to wonder how powerful a factor performance might be in effective teaching, knowing that personality characteristics like those we saw are not easily acquired. However, in many cases, we noted that even 'average' teachers showed an unpraised, more-achievable kind of strength — a sort of quiet valor amidst confusion, conflicting demands and absence of recognition. Not all teachers may be able to demonstrate dramatic flair or creative intellect. But some showed unusual emotional sensitivity and dedication to their jobs, particularly to children. All of them are dealing on the frontlines with many of the problems the society finds very difficult: broken families, increasing criminality, a fragmented sense of national identity. They seemed often, to us, heroic. Witness this teacher's remark, noted during the course of the Study:

> I have a girl who chose to come to me. She said she was going to kill her father. She hated him . . . . I called her mother from school and asked if the girl could live with her. The girl ran away from the father's house . . . . Two weeks later the mother called and thanked me. It was really nice that she did that. So many times you do things and never know what happens.

Or this one:

> I'm blunt. I will not hold back that I feel the family is responsible for the student's attitude and behavior. I ask: 'Do you hug your kid? Do you love your kid?' As far as I'm concerned, that *is* my business. They (parents) are difficult by withholding information, by being unwilling to help me with support by caring for their kid, by not being accountable for the kids.

These are not the problems teachers have been trained to work with — nor do they often receive recognition for grappling with them. But these kinds of demands and issues are unavoidable when working with young people today, and they have contributed to a confusion of educational mission among teachers. One female gym teacher confronted a researcher and asked: 'What am I to do? Prepare them for life? Teach them skills? Build their bodies? Pick up the pieces?' The teacher recalled a distressing incident that had occurred to her in the previous year: a student had asked, in tears, to be released from class because she had had an abortion the day before.

Again, disturbing teacher voices:

> They [the parents] are just crazy. Both the parents and kids have a really psychologically horrible environment at home. The parents play out their problems with their kids. The father of one of the boys who was just in here went out into the garage and killed himself. One half hour before, he told his son that he was disappointed in the son's performance. The son will never get over it. If you wonder why the kids are not doing well in school, take a look at what's happening at home, and then you realize it's a wonder these kids can even get to school.

> I talk extensively with kids about problems; but not often with the parents. Parents sometimes do bring it up. One mother came in and told me: 'I'm dying of cancer'. Another told me my student's older sister was having mental problems . . . . But I never bring it up, never ask them anything.

> I'm usually sending home complimentary notes. That's not covered on our forms. I give the kids envelopes, and tell them, 'Address this envelope to anyone you want to have good news about you'. They are often addressed to the parent they are not living with if the parents are divorced. I have them put their name on, too, so they can open it if they want.

A surprising number of teachers took time to write these kinds of comments, unsolicited, on the Study's questionnaires. We have no way of knowing how many teachers had similar feelings and experiences, but didn't write about them, or even think of writing about them or telling us. We came away believing we had spotted the tip of an iceberg, wishing we had a better idea of the frequency with which such problems reach the teacher's desk, and wondering how often they try to deal with them. Those teachers who touch children's lives in these dimensions must be affected deeply, and we can only dimly fathom the ripple of effects on the academic side of the school program.

A teacher who accepts the challenge of trying to ameliorate troublesome personal difficulties of students runs risks. Sometimes a parent resents what he or she considers to be teacher interference — even if parents, as we also found in the Study, generally seem to want schools to help with such problems. Inevitably, a teacher's action is sometimes ineffective, or even counterproductive.

What motivates teachers to become involved? Why do they keep at it? According to the first teacher we described, Mr. Fasman:

> One of the things that's helped me is that I've never forgotten the grief of being a teenager. I keep reminding myself when I was a freshman, I didn't remember where my locker was. I keep telling myself that when freshmen come here, they don't know whether their locker is in D-wing or E-wing. They forget their lunch, and it screws up their whole day. *Little* things throw them. And I *remember* that. I keep forcing myself to *remember* what it was like when I was 14, 15, 16, 17 and 18. I don't ever want to forget! Kids come with resistance. You have to involve them. If you don't, you never pull them out of their own teenage world, a world made up of kids.

Mr. Fasman is one of the exceptional teachers. What can be done to revitalize 'ordinary' teachers? Eisner and his colleagues interviewed teachers who said they sometimes went a whole day without speaking to another adult. Their discretionary time is very limited. How can an environment be created that allows teachers to learn, on-the-job, about the quality (and the subtle nuances) of their teaching in a form that is ongoing and constructive? The evidence suggests that teachers are more than willing, indeed eager, to receive positive assistance. According to Liora Bresler:

> The teachers we spoke with were eager to get feedback, to feel that somebody is interested in what they are doing. The mere fact of our presence often had the effect of motivating teachers to come up with new ideas, e.g. instead of the usual final exam, assigning students a presentation in class on any relevant subject of interest to them, and bringing magazines and different materials to class to motivate the students to read. Not only did they try to improve the quality of their teaching, but some teachers even began to dress more attractively.

Superintendent Paul Sakamoto, sitting in on early reports in the study about teacher isolation, took matters into his own hands. When he heard that teachers have little opportunity to interact with other teachers to receive feedback or encouragement from peers, he made a point of visiting every teacher in the district for one full class period to observe teaching. He repeated the effort in succeeding years.

In response to a newspaper interviewer, he said, 'It's been extremely, extremely important and a good use of my time. It's given me a sense of what's happening in the district, and it's also helped me make individual or personal contact with teachers.' Undoubtedly, also, teachers began to realize that the district cared about what was happening in the classroom. To further relieve teacher isolation, Sakamoto has asked department faculties within high schools to visit each other in the classroom. 'How else can they assess what's really going on in the department?' he asks.

It is tempting to look at the Mr. Fasmans, the Mrs. Mufflers, Mr. Hills and Mrs. Hainsworths of the classroom, and to demand their performance as a standard. These teachers were exceptional, to be sure, but exceptionally able teachers were evident throughout the Study, both as professionals being observed and as colleagues of the Stanford-based researchers collecting data and helping to analyze it. Is that expectation reasonable? The issue of *Update* that included the reports about their skills also cited another teacher, history teacher Roger Sands [fictitious name], a veteran teacher who is running out of steam. The researcher who shadowed him wrote:

> Roger Sands is a worthy teacher, a good-value-for-your-tax-dollar teacher; even the kids will tell you that. Moreover, he truly cares about his students and holds out much promise for them.

But Sands has worked for years without recognition or help, and his technique is wearing thin.

> What Roger Sands *really* needs most is validation — to be told he is a competent professional — and he needs to be told this by people who have distinguished themselves in an educational hierarchy he both acknowledges and respects. Surely this is not too much to ask.

Let us be sure to add that principals and district leaders matched teacher commitment with their own high degree of dedication and capability. Professor Sanford Dornbusch worked closely with six principals on his 'Families and Schools' project. In his words: 'Each is smart, hard-working, and cares deeply about the educational mission. . . . they are an impressive group — one that any industry or occupation would be proud of.' Noting their commitment, he added:

When a principal *didn't* show up at a meeting, it wasn't an insubstantial crisis that kept them away. It was usually more on the scale of a teacher's wife committing suicide [actually happened]. What you or I consider a crisis is routine for them — like a kid getting arrested. There's *always* a kid getting arrested for them. At a typical meeting of six principals, four or five will be able to attend. Only once was there less than four principals. They knocked themselves out to come.

Let his remarks stand for a host of other administrators with whom we worked in the course of the Study.

These are tales of high school valor, certainly. We have tried to show that despite the odds, there is still a great deal about which the public can take pride in today's schools. Let Barbara Porro, another of Eisner's colleagues, underscore the point as she tried to evaluate the overall effects of the schooling she observed:

One of the most surprising aspects of life [at this particular high school] was the feeling I got that teachers and students were 'on the same side'. Regardless of which schooling game they played, student–teacher relationships were warm and genuine. It was almost as if there were two levels of experience running tandem — a formal schooling dimension and an interpersonal dimension. While the schooling experience was often ineffective, the human dimension was fully functional. Maintaining positive relationships was a priority for both students and teachers. For example, there was rarely any need for teachers to discipline or reprimand students. The tone of the class was informal, friendly. Students and teachers laughed at the same jokes. There was a feeling of mutual respect. Even when students rejected the curriculum, they were polite about it. Teachers accepted the students as they are, rather than demand that they become something else. There was no need for students to lie about what they were doing. There were no negative consequences for getting the wrong answer. Teachers respected the rights of students to work in school or not. They did not set themselves up as the adversary. They did not use fear to control. They were simply there to help students get through school. Students and teachers were working together to accomplish the same goal. Their relationships were honest. Their integrity was intact. The contrast between the way school looks — barren, cold, unkept — and the 'feel' of the place clearly indicates that the human experience is most valued here. . . . I do not believe it is possible to educate without genuinely caring about those we seek to educate. The fact that relations between students and teachers are based on mutual respect rather than fear may be the most promising aspect of [this] schooling experience. That people are not going through the motions of being human is a significant step in the right direction.

Chapter 10

The Students

As we read the reports that were taking shape from the various projects, and as we discussed those reports with teachers, school administrators and the Stanford researchers, we saw ourselves developing views of today's young people in high schools that were distinct from the pictures we were getting of them as students. Our aim in this chapter is to convey some of those impressions. Unavoidably and necessarily, those who concern themselves with education cannot completely disentangle the two perspectives.

We have highlighted the observations that follow partly in terms of differences and similarities between what we think we learned about young people in this Study and our recollections of our own adolescence, both because these comparisons turned out to be inevitable for us (we were not studying another species and could not divorce our own lives from those we glimpsed) and also because we came to think that the reader would fall into the same frame of mind.

One of the first things we noted was that our characterizations of some teenagers in the Study generated extraordinary interest among many of the people following our activities, and sometimes seemed to startle them. For example, when the staff distributed a press release with profiles of three students — the students called Bob, Sandy and John in this chapter (all student names have been changed) — phone calls came in from journalists at newspapers, TV studios and radio stations all over the country. Professor Eisner, from whose shadowing studies the vignettes were drawn (and himself no stranger to publicity), labeled the reaction the biggest one he had ever had to a press release. One local newspaper, in particular, insisted (in vain) on knowing which high school the low-achieving 'Bob' attended. (We will say more about Bob later in the chapter.)

The reporter missed our point, of course. Though 'Bob' is a real person, he is not the 'failure' of a particular school or a particular district. We thought of him as a surrogate for students anywhere. Furthermore, teenagers like Bob were around during our own childhoods, and we were aware of them. They were in our neighborhoods; we saw them almost every day. 'Bobs' may even have been in our own crowds. At that time, however, we were not trying to figure out what the schools might do to serve them better. As nearly as we can tell, the educational system was no more capable of coping with such young people, then than now.

We also saw some differences in the students, or think we did. There seem to be more children doing poorly in academic subjects than we recall, but then we have to remember that schools today serve a far greater percentage of teenagers than they ever have. Yesterday's dropouts are many of today's students. An additional difference, though also impressionistic, is that more students seem to be working after school, or in 'release-time' work-experience classes, than in our day. The number of students whose first language is not English is also a striking change, a transformation perhaps more notable in our Study than in schools outside California. But then we have to remember that in the earlier days of the century and the final days of the last, well before we went to school, high-schools were similarly expected to serve large numbers of young people who were most comfortable in another tongue.

The 'Crowd'

Now, as always, high-school students' educational experiences are shaped largely by the other teenagers with whom they associate and for whom they have the greatest admiration, and by the values and interests developed at home. With respect to their peers, this is what English teacher Charlotte Krepismann has to say:

> Teenagers value their friends more than anything else that happens in high school. The observer sees it in the classroom, in the halls where the chattering reaches amazing levels of release, when they no longer have to sit and perform and 'be good'.

These friendships shape the students' self-perception, perhaps even the 'track' that a teenager chooses (or, more accurately, drifts into at some time before or during the sophomore or junior year). A student's expectations are shaped by peers and family, and rarely does a high-achiever associate with a group of low-achievers, or vice-versa.

This track, in turn, powerfully solidifies the total extent of the teenager's associations, day in and day out. In general, a child whose track is 'vocational' associates primarily with similar students, in similar classes. For low-achievers, none of the people they know best may study hard or value higher grades. High-achievers, on the other hand, avidly compare GPAs, test scores and homework assignments. Both groups program their expectations to a major degree on the basis of the values of the young people toward whom they gravitate. In every school, there are dozens of sub-groups of students, who share opinions and shape each other's outlook toward the rest of the world.

Here is part of a description of the dedicated low-achievers that emerged from Eisner's work:

> The low-achievers are professional about doing the minimum amount of work in order to graduate. They often accumulate credits in subjects such as IWE (Inside Work Experience), which commonly involves routine work (checking attendance, throwing balls in gym, sitting in the office and making sure that nothing is 'ripped off') and does not require any intellectual activity . . .

Another low-achiever was described this way: 'Mary was involved, animated and talkative in the non-academic [classes], sullen, quiet and detached in the academic'.

These cliques of students identify themselves to each other by the way they dress, the classes they attend, the way they participate in class and even where they spend their free time. In one of the Eisner shadowing studies, students identified themselves by where they 'hang out' during their lunch hours and breaks — the 'side-lawners', for example, a group of students who have a reputation for being tough and for showing up on the school's side-lawn whenever they have the opportunity (including when they were cutting classes). There, they shared food, bummed cigarettes, played cards, swapped stories, flirted, joked and generally had a good time. Said Barbara Porro: 'If you ask the side-lawners what they are about, you will find with a high degree of consensus that they are seriously engaged in getting through high school by doing the least amount of academic work necessary to pass their courses'.

In this particular school, there were other subgroups for the academically inclined, who met in the library and for the athletically inclined. At another school, different cliques met at different fast-food restaurants for lunch. The need for this self-identification and peer-contact is so strong that students pay for fast-food lunches even when their low-income family backgrounds entitle them to free food passes in the school cafeteria.

Researcher Sandra Schecter shadowed a young Hispanic woman, who described a school sub-group known as 'Cholos'. The Cholo women wear tight jeans, high heels and much make-up. The men sport dark, overgrown brush-cuts.

> In the world according to Lilia, there are good girls and bad girls, good boys and bad boys. Lilia was 'bad' two years ago, when she was in seventh grade and she is determined never to be bad again. 'What is bad, Lilia?' Well, for starters, the Cholos are bad. She used to hang out with them in seventh grade. 'They'd say, " Let's go smoke some weed. . . or let's cut school", and I'd do it just to be with the group'. Lilia is uncharacteristically harsh on the subject of Cholo girls. Cholos are 'bad. . . they go along with anybody just to be in the group. . . They don't think of what's gonna come ahead'. For example, one Cholo girl she knows 'is gonna try to get pregnant from some guy; it's like a bet . . . it's not as if they sit down and they think, "Okay, if I have a baby I'm going to have to leave school, get a job and take care of the kid for the rest of my life until he's eighteen"'.

The case of Lilia is an interesting one, for she is bright, mathematically inclined, ambitious and cooperative — yet several of her courses are non-academic, like home economics and typing. Moreover, her friendship with two other low-achieving minority students, Maria and Amie, is exclusive and possessive.

> The girls negotiate the throngs [before school] assertively. They move in a single unit, keeping one another in tow. Now one, now another, stops to greet a special acquaintance by name, but the extra-peer group discourse is not prolongated. . . . The girls are unreservedly warm with each other. They hold hands, place supportive arms around one another's shoulders, seize the other's wrist with the frenzied grasp of a drowning person at the sighting of an attractive member of the opposite sex. . . . Lupe and her friends will reconnoiter again at brunch-time, at lunch and after school to update each other with late breaking news bulletins on the state of family affairs, problems with class assignments and skirmishes with members of the opposite sex.

The girls are inseparable: 'We don't hang around with anyone else. It's just us three'.

The school's counselor predicted that Lilia would shed these friends as she up-graded her courses, identifying herself more with a new sub-group of more-academically oriented teenagers. But the researcher was not so sure — in view of the overriding importance of friendships in the Latino community, in view of Lilia's lack of adult academic guidance, in view of the fact that each counselor these days is expected to serve about 600 students, in view of the fact that almost all the counselor's time is captured by problems that demand more urgent attention, in view of the fact that a student like Lilia wouldn't think of approaching one for help. 'Guidance is where you go when you're in trouble', she says.

Different Worlds

Generally, the most-widely publicized calls for educational reform are heard from those adults whose academic experience has been favorable. Reports that teenagers today read poorly, or that test scores are dropping, or that children don't know whether Hiroshima occurred before or after Pearl Harbor, start them wondering what has become of the classrooms they remember. For the most part, however, the educational path followed by most of today's arti-culate critics is still intact. Visit any Advanced Placement class. For those who fear that schools have become undirected, or entirely custodial in function, as apparently many do, the following observation of college-bound high school students drawn from the Study should prove both reassuring and familiar:

> Teachers seem highly aware of their responsibilities for preparing students to succeed in college. Top track students are rather sheltered and tend to continually get the best teachers. These classes function essentially as college courses in terms of curriculum (college-level content) and course design, the lecture being the most frequent mode of instruction. Here the emphasis is on delivering an established body of information. Homework is heavily emphasized. Students know well the rules of the game — what they are expected to do: bring pencils and paper, take notes, turn in homework.

Here is an excerpt from one case study:

> They [students] were writing as fast as they could and seldom had
> time to look up and simply absorb what the teacher was saying or
> think about what it meant. When the teacher is not lecturing formally
> they might just sit down and 'listen politely', but if the discussion
> happened to turn to something relevant to the information learned in
> class or an upcoming assignment, they could be seen suddenly
> whipping their notebooks back out of the bike bags and quickly
> scribbling down information.

Compare this with the story of 'Bob', to whom we referred earlier in this
chapter, an intelligent 'side-lawner' who, during Porro's two weeks of
observation, carried no books, seldom had paper, rarely read or wrote (except
numbers in math class) and almost never knew what the teacher was talking
about. During Porro's eight-day observation, Bob cut six classes, was late to
five and was suspended for getting into a fight. (Suspension from school,
ostensibly a disciplinary measure, worked, he thought, to his advantage; he
was able to miss two full days of classes with excused absences, during which
time he earned $30 mowing lawns.) Porro chronicled Bob's novel methods for
cutting classes:

> One day when we arrived late to class, we found a note on the door:
> 'In library, 7th period.' Bob ripped the note from the door and stuffed
> it into his pocket, explaining, 'Here's another way to cut class. When
> she says, "Didn't you read the note that we were meeting in the
> library?" I'll say, "What note?"'

Said Porro of this particular clique of students: 'You may find this difficult to
believe, as I did at first. You may wish to call him the exception. But, in fact,
Bob is one of many students I observed who is opting out of learning and is
getting away with it'. She asked one student if he wanted to graduate. 'I don't
want to graduate, but I'm *going* to graduate', he replied.

These students had a view of life in which there was little connection
between classroom-centered activity and anything else. They were sure that
they would outsmart life, as they had outsmarted high school rules. Bob even
thought he was going on to college. When asked why he was so skillful at
getting away without working, his answer was, typically, cheerfully cocky:
'Because I am awesome, and the teachers are lame'.

Bob does not represent an isolated case. The Study chronicled others. A similar student attitude may lie behind this response, written by a student who was peeved at being asked to fill out a Study questionnaire:

> It was a real waste of time because it took time out of my auto shop class and I could have finished painting my engine by now, but thanks to this survey I have been delayed even further in the process of rebuilding my Chevy 350 4 bolt main engine for radical street use. Yes, soon I will be tearing up the streets, lighting up the tires and powersliding around corners and generally wreaking havoc on the whole town while having one helluva good time. Also, I won't be thinking about school at all because it will be summertime and being a normal teenager I will have better things on my mind (Heh, Heh)!

Student Passivity

Bob is unlike most other students in the cheerfully active way he works to sabotage the school's attempt to provide his own education. And his ingenuity may even translate into a wily success when he enters into the full privileges of adulthood, given his engaging personality. Most students we observed, however, are simply too passive and accepting for Bob's kind of inventiveness — whether used for or against their own education. Typically, they view a diploma as a ticket to the future, without much thought about what the diploma is supposed to signify; many feel school is supposed to 'act' upon them, or perhaps anoint them, that they will somehow absorb knowledge without any active participation or involvement.

One teacher-researcher interviewed a finger-snapping, gum-popping student about her courses, and concluded that she was typical in her vulnerability, her compliant and unquestioning attitudes, and her craving for attention at any almost any cost (even playing the 'airhead' in classes). She rarely had an idea, or even a reaction, about her class-work without prompting. She passively agreed with the interviewer. Here is an excerpt from the interview:

> Q: Do you think most kids do their homework?
> A: It depends on the parents. In Spanish, you work hard if you want a good grade or at a different pace if you're there because your parents think you should take that class.

Q: Does the student have a responsibility for his or her own learning?

A: Yeah. It has to be. But sometimes kids won't do homework unless their parents push.

Q: On the average day, do you come away feeling that it's been a good day, that you enjoyed school?

A: Yeah, today it has been because we didn't sit there just doing nothing particularly interesting.

Q: What makes a class particularly interesting?

A: Like in US History, we saw a movie and had the whole class join in and give our viewpoints. That's what we usually do. In Spanish, we had games and it was fun . . .

Q: Are there any classes that you dislike? What I'm really looking for is what makes the difference between a class that you dislike and a class that you like.

A: Hmmm. English, sort of, because I'm not that good in English. Some parts I can do fine, but when she gives us a topic for in-class writings, it's really hard for me to write on something I don't know about or if it's a weird topic. Unless I get suggestions, then I wonder where I start.

Q: Do you like to read?

A: No . . .

Q: Okay, in terms of your future, are there some classes that seem more meaningful — either for college or a career?

A: Uh, English, a little, because I'm probably gonna be a (names profession) — or biology because I'm also thinking about being a (name profession), and you have to know the terms and stuff. It depends on what happens . . .

Q: The only class that seemed relatively peaceful was typing. Do you think of that as a pretty relaxing class, or is it just to get credit?

A: Yeah. I'm there mostly to get credits. But I have to know typing, to do business letters and type for college. It's fun and I like it as my best class.

Q: It seems like you have a lot of freedom in there, but not much interaction with the students.

A: Well, we ask questions sometimes. Yeah, it's neat. Everyone seems to have a purpose. I wish it were always like that in English . . .

The image that emerges is one of teenagers psychologically removed from the educational process, passively accepting schooling and its limitations. They don't truly comprehend the connections between what they learn in class and how it will apply to their future, in the world outside the classroom. They accept it all as part of going to 'school'.

The Messages They Get

Adolescence is a contradiction: clearly, people this age are in the throes of early adulthood, yet retain the vulnerability and perceptiveness of children. As described in an earlier chapter, they are adroit at picking up the inconsistencies in our educational world. More than most adults, we think, they are likely to believe what they see rather than what they are told. Adults assert that education is important, but such a statement frequently seems at odds with their own observations. Here is part of a report from David Flinders, another of Eisner's colleagues:

> Student interview responses frequently echoed teacher concerns regarding their schools' lack of support and financial resources. Students repeatedly criticized the elimination of humanities and vocational course offerings, special activities and sports. Several groups of students complained of old, worn-out textbooks, and of classes where there were too few textbooks to go around. Students repeatedly described their schools as filthy, overcrowded and in dire need of better maintenance. Some students felt that the administration was too lax in responding to requests for repairs: six months to fix panic bars on fire exits, a year to fix broken and defaced lockers, two years to repair toilets. Several groups of students, at both schools we visited, also complained of rats and cockroaches. Almost all of the students we interviewed associated these problems with their schools' meager budget and seemed keenly aware of the limited financial resources available for education.

Students note other messages. Said Bresler:

> Students as well as teachers are quoted as saying that 'music, art and athletics won't get you into college and they won't give you a way to make a living'. The music and art departments are often on the top

floor or in a faraway corner of the school, as opposed to the central location of the 'academics'. Their location in the day, often a sixth or seventh period when the concentration level of students is lower, also reflects their 'extraneous' position. Even subjects such as music are often justified on extrinsic grounds (e.g., fostering a sense of belonging, developing steady working habits and strong values of responsibility, promoting social or job skills, and so forth).

And one of the message they receive is that future jobs, not school, are important. In fact, they are led to believe the latter is only a means to the former. Again, a quotation from the press-release mentioned earlier in this chapter:

> For Sandy, school is just a halfway point to getting a job. [Researcher Sigrun] Gudmundsdottir says that Sandy and her friends frequently talk about their own [after-school] jobs or some wonderful job that a friend has and all the exciting things this friend gets to do. 'If they do not like the job they have, then the discussion is about the kind of job they would like to have', notes Gudmundsdottir. 'Glamour and money are not the only things that make a job good. To my surprise, benefits are considered as important as pay.'. . . The school's business teacher echoes Sandy's attitude: 'My first aim with this course is to get them aware of the world of work because that's where reality is. If you can't pay your rent and bills, you really don't care who is the president of the United States. You want to eat first and take care of your family.'.

Again, the adolescent, with his or her yes-no view of the world, picks up subtle (and some not-so-subtle) divisions: school/education is at one polarity; reality is at the other. Or, as one student expressed blithely on Professor John Krumboltz's questionnaire: 'School is for learning! Parents are for living!' David Flinders picked up another: school is for learning; people are what you remember:

> Their interview responses suggest a marked separation between curriculum content and the students' affective school experience. When asked what students felt they had learned from a particular class, their responses were invariably focused on specific subject matter and narrowly defined academic skills. Generally, these answers were given with much elaboration — 'we've learned about living

organisms', 'the metric system', 'how to work math problems', and 'information for the test'. In contrast, when we asked students what they would *remember* about a given class, responses overwhelmingly centered on the personal characteristics of their teachers and the non-academic aspects of student/teacher relationships. One group of students said they would remember their teacher's face, the way he talks to people and his brightly colored socks. Other students reported that they would remember their teacher's efforts to help a friend find a part-time job, another teacher's sense of humor, and one teacher's understanding toward the student's ambivalence about going on to college. Overall, these responses suggest that, from the students' perspective, curriculum is what you learn, what you will be tested on; people are what you will remember.

Future Fears — Future Hopes

High school students showed a marked degree of apprehension about the future — particularly surprising, perhaps, to those adults who grew up during the seemingly more unstable and tougher days of the Depression or World War II. At the same time, students during the course of the Study exhibited a startling outward self-confidence, a belief that they hold the key to the future, that 'later' will be full of promise for them personally. The fears did not necessarily come from the low-achievers; the hopes did not all come from the high-achievers. We noted that the same students express these contradictory feelings at different moments.

If anything, our studies suggest that students don't understand much about the future or its demands on them. Wrote one student on a questionnaire: 'Where do I go from here? High schools need to help students overcome this fear'.

Perhaps the apparent ignorance of high school students about the future they face is instrumental in their fearful attitude toward it. Most seem to have little contact with adults and, for many, the adult associations that carry greatest intensity are those with their after-school-hours supervisors at work, who are often in jobs with very low status and pay.

Additionally, a picture that emerges from Weiler's research is one of teenagers frightened by their own ignorance. Of 950 students surveyed, 84 per cent said they were anxious about the nuclear arms race, and over 80 per cent were worried about a nuclear war breaking out in their lifetimes. A smaller group of 200 was asked an open-ended question: to name the two or three most important international issues of our time. Fifty-six per cent responded 'nuclear war', with an additional 9 per cent citing war in general. But, as we have reported, their concern was not accompanied by much knowledge of the political world internationally. Said one: 'I would like to know more, but in some ways it seems hopeless. I'm really scared about a nuclear war, but sometimes I'd just rather not think about it, so I don't try to find out more or look into the subject.'

Sandra Schecter wrote that 'students need to be delivered from the humiliation of their own ignorance'. She was actually referring to students' fear of being called on in class — but the comment haunts. Could it not also include, we began to wonder, their limited knowledge of their own abilities; their narrow working knowledge of the world?

Despite these fears, however, many of the students looked forward to graduation, to adulthood, to 'freedom'. Many thought they were going on to college — a greater percentage than the school's track record would suggest. Many dreamed of wonderful careers — careers that would satisfy and interest them in a way school had not, even when future plans were unspecified, or vague or commonplace. Wrote one researcher:

> They live on this gut level in the 'now' with some vague plans for a future that is not terribly clear nor anywhere as safe as right here in these high school halls with the 'grungy' scribbles on the walls, the bits of today's lunch underfoot and marvelous plans for 'later'. They part with one another on this same bright note . . . 'later' . . .

The Students' Future

Many of the students seem to have an uninformed and largely unexamined view of their present circumstances in school, as well as an unreflective conception of their future. They confuse 'education' with 'job preparation', and their view of the latter is narrow and self-absorbed. They do not see themselves as future voters who are responsible for understanding the issues upon which they will vote and the history they will help determine. They don't have the insight that helps them comprehend the relationship between the past and the future: 'Why should history be emphasized when the future is more important?' one student asked. More abstractly, they do not see themselves as guardians or torch-bearers of cultural tradition or civilization. They tended to emphasize what is immediate and what is likely to draw personal income as their biggest priorities. This obsession with the short-term goals surfaced repeatedly (and sometimes inelegantly):

> To me, the basic importance of high school is to prepare you for your entry to the job-finding world. Unless science or history is to be your field, it is irrelevant what the 44th element is. Or when the Battle of Hastings was. It is fun to know, but I'd rather learn to speak well or to deal with others better. . . . Career Decision Making processes should be emphasized more. I also think that an articulate person has a large advantage over a 'just fair' speaker in getting any job or position.

> I think than (*sic*) education is important, but only when you're going to use what you learned for the future and for your job. Things like history, science & advanced math are unimportant to me because I doubt that I will ever need to know them or use them.

> Music interests me, but I do not feel it should be taught in high school.

> When the student is talked to & knows what career they want to go into, why don't you just let us find out the prep. classes we have to take for it & let us prepare for *our* own carrers (*sic*) . . . you tell us, 'you want to be treated like an adult, then act like one'. But makeing (*sic*) me study classes that I have no interest in doesent (*sic*) look like Im given the freedom of choice which is in the constitution isnt it? Well, please concider (*sic*) what Im saying because Ive seen alot of Drop out & unhappy students.

> You might have asked what field of work we would like to go into once out of high school and then ask if we think high school has helped us any with preparing for this job (or preparing us for college).

> Many people today, like myself, are looking for a job that they can keep for quite a while. I feel that they must understand the steps involved in applying for a job and then holding that job. They must have self-confidence in themselves.

Again, Garet found ninth-graders hoping vaguely hoping for a college future, then awakening to reality by their senior year. One student, in large and uneven letters, wrote: 'Schools are hopeless. I get a 'C' average, but still plan on going to Stanford. P. S. You should have a motorcross team'.

Who Listens?

Time and time again, the shadowing studies of Eisner's group revealed the large need by young people for adult attention. One teacher told a researcher that a student had 'blossomed' since she was shadowed by the researcher for two weeks; another student, shadowed by a teacher at the same school in connection with the Study, enrolled in that teacher's class the following year.

Researcher David Grossman, who analyzed student comments for Weiler's world-attitude survey, said that the results were less surprising than the students' response to the questionnaire itself: 'The attention and respect given to their opinions was new to them, and they were flattered. They *did* have opinions, irrespective of their knowledge levels. . . .'

Eisner's researchers point out that, with one-parent families, or both parents working, many students are burdened with household work, or have no after-school supervision. How often do today's teenagers talk to adults about their feelings? Some research suggested that for many, the only adult attention they get is at low-echelon work-study jobs, working as clerks or cashiers. The interchanges with co-workers, who are often lacking in education themselves, may be less than stimulating. Some, on the margins of a questionnaire, issued cries for help:

> How many [teachers] talk to you as an *individual*? More on an emotional level. I can see a lot of people around me dying emotionally because everyone's draining them so much. It's terrible. There's a lot that goes on inside that schools could help with. Relationships are important — people get strange ideas about them from TV and all. I think someone should put people straight. The bell rang. Thanks for listening!

Professor Sanford Dornbusch, who studied the relationships among students, adults and grades, found that adult contact could be important: students who engage in more adult-supported or adult-controlled activities tend to have higher grades; so do students who would side with parents when their opinions conflicts with peers. He found that 'the low-key use of praise, encouragement and offers of help is associated with higher grades'. Parental participation in school functions also has a 'substantial positive association with grades'.

Such observations raise other questions. How much was the past 'success' of schools due to the influence of a mother actively affirming values and checking up on homework? What impetus do children have to succeed at schoolwork besides the fear of classroom humiliation? Is there anyone to assert the value of knowledge and understanding (not just the amorphous commodity of 'education') in their lives?

Given that the one-parent family is a new reality in American life on the scale we see it today, we became interested in what might be done in schools to ameliorate some of the resulting problems. Can schools help meet students' needs for adult communication? And what can teachers and school administrators do to provide assistance to single or working parents?

Professor Dornbusch worked closely with principals to probe such questions. One school, as a result of Study-based discussions, accommodated working parents by making any needed phone calls to the home at night. Another school, headed by Principal Samuel Johnson, made an extra effort to get parents involved in school activities. Dornbusch suggested that working parents' phone calls home from their children might have a positive influence.

There are occasionally 'simple' solutions to complex problems — but some of them can be found only when informed participants are able to find the time to gain the perspective to think of them. Dornbusch found that many parents are willing to search for solutions once they understand the nature of the problems. Good news, too, emerged from the teachers. Said one:

> It seems that for American parents . . . the higher the education of the parents, the more interested they are in their child's education, but it's not true for us (in the ESL program). We have a very big group of very uneducated parents who are very active. Someone is always here representing the family, usually the mother or the father of a family with a lot of siblings. They have a big stake in having their family do well . . . . Our parents tend to be very supportive of us. They often come from a country where the school is always right.

There is lots of scope for 'solutions', since the problems are so serious. One teacher noted:

> Our kids in our department, I don't know how they do it. Some work forty hours a week on swing shift. Then they come home and

do homework. . . . They don't have a lot of time to sit or relax. . . . and they have to study so much longer because English is not their first language. It's real hard at home, taking care of cooking, cleaning and younger siblings when both parents work. . . . These kids don't usually work to earn money for themselves. Most of the time they don't even keep it, they hand it right over to their parents.

More Changes

One of the pronounced changes in high schools during recent decades, as has been remarked here repeatedly, is the much greater percentage of American teenagers staying in school. Many of them have overwhelming responsibilities at home and at jobs, and in an earlier era would not have had the additional obligation of going to school:

Lilia has the responsibility of seeing that the domestic needs of her household are met. She puts it less euphemistically: 'I have to do everything at home'. The deixis of 'everything' eludes me, so I ask Lilia to expand on her home routine: 'I wake up at six o'clock . . . I take a shower . . . I have to brush my sister's hair . . . I have to wake up my big brother . . . I wake up all of them . . . [my brother] goes out and warms up the car . . . we take my sisters to [elementary] school and then we come over here [to school]. After school . . . I go down to El Camino, wait there until three o'clock . . . for my brother. Sometimes I go [grocery] shopping for my family . . . and then when I get home I do the dishes or any housecleaning to do . . . or sometimes I just do my homework and then do the housework . . . I cook for my dad and my brothers, too . . . They want to eat so I feed them all . . . posole, tamales, enchiladas — I can make everything . . . Then [in the evening] I have to iron my big brother's clothes and make his lunch . . . '

After school every day, Sandy goes home, has lunch, watches the soaps, does her share of the housework until it is time to go to work. She works at Sears almost every night from 6 to 9 p. m. and arrives home around 10 p. m. She does very little homework, generally after work or early in the morning before school.

After school, [John] works three-and-a-half hours every day as a cashier at the family bakery — a job that brings him the things he really values. His extra income buys him prestige: jeans, a Jordache jacket, a perm and visits to the Kung Fu movies with friends. John guards his remaining evening hours jealously. . . . Homework is generally not part of the agenda.

Dornbusch found that working twenty hours or more a week at a job was associated with lower grades, especially for students under age 16.

Additionally, today's classroom, especially in California, is likely to contain an unfamiliar ethnic mix. Take, for example, the 950 students Professor Weiler surveyed as part of his study: 30 per cent were non-Hispanic whites, with 29 per cent Asian-American, 14 per cent refugees from the Far East and Central America, 13 per cent black and Native American, and 11 per cent Hispanic. Over a third were born outside the United States, and just over half spoke English as a first language. This mix may not be representative of the US as a whole — on the other hand, it may be a portent of the future.

The repercussions of this diversity can be staggering. The system, as we've suggested, can function for a time on its momentum. It can maintain a basic, day-to-day level of competence that seems little changed year in and year out. But the rapidity of demographic change demands adjustments at a more accelerated level than schools are accustomed to: scheduling more ESL and remedial English classes, finding teachers who are competent to teach the subject, accommodating to different parental responses to children's experiences in schools.

But it means more of an adjustment at less tangible levels, too. It means classrooms of students whose educational expectations may be very different from the teacher's, and from the school's as an institution. Even the conventions for acceptable social behavior may be unfamiliar. Such changes may affect what courses students value, how much they value the educational process itself and what they expect from it. It affects how much energy they are willing to invest in extra-curricular activities that are part of the school's culture, perhaps even part of its curriculum — like football, or band — but perhaps not part of the culture from which the children come.

Chapter 11

Consequences for Today's Schools

It may be the case, as many people have come to believe, that schools are beset with so many profound and intertwined difficulties that nothing can be fixed unless everything is fixed. Our Study certainly provides plenty of evidence to support this view, and we can only be heartened by the many serious attempts initiated in recent years to approach educational reform on a broad front: from providing adequate finances, to undertaking basic curriculum revision, to providing career advancement for the very best teachers, to launching ambitious and well-conceived attempts to alter traditional patterns of teacher selection and education.

Nevertheless, the Study we chose to design was rooted in the schools in our part of California, as they are today. We might have decided instead to construct a more visionary picture of secondary education, one toward which we and our neighbors in the schools could all join forces and strive to achieve. But we did not take that path, for several reasons. For one, we wanted Stanford to be engaged with the realities of life in classrooms today; by such an approach, we thought, university-level scholarship would be enriched by attempts to understand actual conditions, and those who work in schools would benefit from different analytical perspectives on the circumstances they have no choice but to face. For another, we started with a conviction that the current state of affairs did not arise capriciously, and that for progress to occur it is necessary to understand present circumstances, and move on from there.

We started also with the bias stated at the outset: that those in schools have the intelligence and commitment to take the major steps in resolving the difficult problems they face, if given help and other forms of support, both steadily and strategically. We continually found evidence to reinforce this view, and the corroboration further solidified our intention to base the Study on schools here and now.

As a consequence, the pictures we have developed about schools, and that we have tried to convey in the preceding chapters, are shaped strongly by their current problems. The resulting portrayal is not always flattering. We still believe, however, that the pictures present the kind of fidelity to actual schools and classrooms that is essential for progress. And now, in this chapter, we will try to step back a bit from the detail presented earlier, to present a wider angle on the educational problems we have highlighted.

Our Own Lenses

Before trying to describe a general perspective on educational conditions in the high schools we came to know, it is worth reminding ourselves, as does Professor Robert Calfee, that,

> . . . We tend to romanticize the past. [Our] recent studies conclude that "for many young Americans formal education has become a ho-hum operation". [However] this fits with my recollections of high school more years ago than I like to recall.

A younger researcher, graduate student Sandra Schecter, thumbed through a high-school algebra textbook during an all-too-familiar classroom session:

> The Algebra text, appropriately entitled *Algebra*, is similar to the one I used when I was in Grade 9, some twenty years ago and at the diagonally opposite end of the continent. In fact, it may well be an updated edition of the same text. I mention this not by way of criticism of the text or even the school for using it, only that it is incongruous to me that given the consistency of formal math curriculum over the years teachers are not by now more adept at finding ways to make the concepts accessible to students.

The way teachers teach, the textbooks used, the various classroom characters taking their various roles — all have the disturbing familiarity of something long-forgotten, but suddenly re-seen through adult eyes.

In reading reports on educational reform during the past decade, and in reviewing our own observations, we have tended to assume that the observer is a constant figure, rather than a person influenced inevitably and steadily by his or her changing times and shifting expectations. Many of the classrooms James Conant saw were as barren and stilted, surely, as the ones in many schools today. Decker Walker tells us that while many courses on the periphery of the school curriculum change, in the end they leave only a residue in the high school, while the core program remains surprisingly constant. Conant was not highly critical of what he saw. He might, however, express greater unhappiness with today's science programs than he did with those he wrote about thirty years ago. To what extent are schools different, and to what extent do we now tend to listen to critics with different preconceptions?

In the Study, we were often reminded of the subjectivity of the researcher/observer. As Professor Elliot Eisner noted: 'It's interesting that we talk about ''findings'' as if we go out and discover something rather than recreate our world'.

We did not always recognize the world we had re-created — but occasionally we had flashes. Researchers tended to be dismayed by knowledge levels of students (and even of teachers, in some cases), forgetting that university researchers are themselves largely the successes of the educational system. Typically, the researchers were the students who were sufficiently motivated to graduate from high school, go on through college, return for an additional degree or two and then teach others; a little candor and memory led us to recall how uncomprehending or unconcerned we once were by the educational indifference of some own high school peers. The world-view of an underachieving student, bored in class and neglectful of homework, was always a puzzle, but we all remember the low-achievers (perhaps as much as a quarter of the class): their interminable, humiliated pause after a teacher's direct question, the very low levels of engagement revealed by their class presentations.

Now that compulsory classroom attendance keeps more of these low-achievers in the classroom, we must temper our judgments if we see a new generation of teachers no more able to reach these students on the academic margins than their predecessors were, or if they find themselves unable to cope with the sudden increase in their numbers. The distress a few researchers expressed about the apparent lack of effectiveness of some of the high school teachers shows again our own bias, at once an increased objectivity and subjectivity. The private crises of individual students that we had viewed as 'normal' during our own teenage years appear more ominous today, given what we now know of the future of these children.

The teacher, once viewed as an all-powerful authority figure in charge of the grade-book, is now a fallible adult, trying to manage a classroom of semi-adults, without much encouragement. Nowadays, unfortunately, there is often the temptation to see the student as a victim and the teacher as an incompetent — a point of view that has surfaced in many popular reports of the 1980s. What we preferred to see, and then saw, of course, was a teacher who was overburdened, underpaid and discouraged, while trying to cope with a difficult job that had become more administrative in recent years. We saw them working hard to prepare students to pass State- or college-determined examinations, while also serving as psychiatrist/baby-sitter to teenagers they do not have the time or training really to understand.

The Breaking Point

Like others who have tried to look at schools sympathetically, we found that our high schools are staggering under a social load they were never meant to carry. Furthermore, they are losing ground, and probably will continue to slip, as our educational system is subjected to new tensions: the biggest immigration wave ever to hit the United States (which has, in some schools, reduced the native-English speakers to a quarter of the class); increasing poverty (particularly in single-mother households); enormous shifts in the job market, which makes the future that teachers are helping students to meet even more problematic than in the past; more-inclusive and broader societal expectations about the role schools should take, coupled with the decreasing beneficial influence of other stabilizing institutions, like the church; and the apparent weakening of the nuclear family, which may create emotional trauma for children and drastically alter the psychological and educational support a child receives outside the schools.

These changes certainly are prominent in the California schools we studied. Judging from the reports of others, they seem an inescapable part of the national scene. Many of them, like the changing nature of American family life, can be addressed productively by the schools with thought and effort. But, unattended, all of them limit severely what teachers and school administrators can do in response to the intense pressures they face.

This picture may seem overblown to many readers, but consider the comments our researchers occasionally received from teachers. This one was written by an English teacher; it can stand for many others:

> I do a lot of listening. Kids can write anything they want to at any time. There are those who kind of write it out — give me a note . . . . Sometimes girls will tell me they are molested by step-

fathers. A girl came in yesterday and told me she was pregnant. . . [another] girl came in and told me incest was going on in her home, and she was a victim. . . . I do an enormous amount of work on self-image, relationships and personal power. It's my hope with those added strengths they'll be able to cope. They're all beautiful kids.

Where Are We Going?

In principle, the educational system can be tailored to accommodate to the changes in our society, but even accommodation becomes unlikely, as well as difficult, without consensus about the direction in which that society is moving. Public schools cannot be expected to solve the Nation's problems without significant support from other institutions, and that support is unlikely to the degree necessary unless schools are viewed as promoting national goals. Nor can tax-supported institutions risk floating an independent vision of a desired future. Neither can they, for long, solely promulgate a nostalgic, idealized version of a simpler past. Instead, they have little choice but to grapple as best they can with the perplexing changes — but without clear indications of where the changes lead, or even where the public wants them to lead.

For schools to work, they must reflect, at least implicitly, where we are we going as a people. We expect our schools to give our students a direction, a purpose, but our schools are a mirror on the diversity and confusion in our culture as a whole. What can schools do to teach Americanism if we aren't reasonably sure what it means to be an American? What values should be learned through the conscious efforts of teachers and school administrators? And the fundamental question: Can we expect our schools to continue to foster ideas about America and its future that the adult society may no longer believe, or at least acts as though it no longer believes?

To a large extent, the approach in schools today to, say, American history is not significantly dissimilar to that used in classes of three or four decades ago. But though the schools intone the same gospel, the congregation has changed. We suspect that one must figure out how to teach the fundamentals of Western culture, say Shakespeare, to a classroom of first-generation Mexicans or Chinese in a somewhat different fashion than to third-generation descendants of Western Europeans. And we must figure out much more: how

Mexican and Chinese heritages become part of the American, and how to teach about current and controversial American political cross-currents to children in high schools who are Cuban, Vietnamese, Salvodorans and Nicaraguans, for example.

Not only is the congregation different, but it is larger and more inclusive. As we have said repeatedly, and as Professor Robert Calfee recalls: 'When I started first grade, around 1940, the dropout rate from high school was about 75 to 90 per cent. By the time I graduated, in the summer of '51, the dropout rate had declined to 50 per cent. By the time my child graduated, the dropout rate was about 20 to 25 per cent.'

In other words, the dropouts of yesterday are in the classrooms today. Societal belief in the value of the diploma as a credential, rather than any real commitment to education, keeps these adolescents in our high schools. It is sometimes difficult to be sure how these low-achieving students are benefitting from the education they now receive in school. Our evidence suggests that they certainly do not get what we hope they will — and maybe they do not get much at all. Still, in the association with dedicated adults and in the contacts, however limited, with other young people, they probably derive more from the educational experience than our impressions suggest, and almost certainly obtain more of value than if they were not in school. We acknowledge that some people question this last assertion, but we reaffirm it.

It's No One's Fault

After more than three years of school–university collaboration to understand and improve high schools, with projects examining the role of school administrators, school teachers, university researchers, politicians, parents and students, we found no villains. Everyone means well; most of the players involved, whether at the school, district, university or State level, are committed and even idealistic, in the very best sense. Many are uncommonly able. But circumstances constantly undercut the ability of each of the influential parties to fulfill his or her own best intentions. For example: the teacher does not have the time or skills to implement the best educational plan, and with startling frequency, is forced to teach a subject outside his or her field; the school administrator often does not have the funds to provide teachers with the training or materials they may need, or even the time to devote to anything more than day-to-day troubleshooting; and State-level policy-makers enact legislation whose effects are sometimes the opposite of what they had intended, not because they are mean-spirited or thick-headed, but because they often have little understanding of how schools actually function.

As Professor Michael Kirst noted in *Who Controls the Schools?*,[1] 'Americans have always believed that education policy is too important to leave to educators. Now, as in the past, an aroused citizenry can powerfully influence school policy'. But when children do not learn, when test scores drop, the blame always seems to drift toward the teachers and school administrators. We have found, and we have tried to document the fact, that this view is oversimple and counter-productive.

No one in the Study wished to overlook or deny the central and unavoidable responsibilities of those who are designated to provide educational services to young people in the schools: not least, the teachers and school administrators themselves. But close collaboration over several years has made very clear the complexities and the mitigating factors for the lack of success we often see.

Education by Accident

During one Study meeting, Professor Kirst likened education to a sick patient, who has been given so many pills and therapies, for so many perceived ills, that it is no longer possible to tell which remedy is working on which symptoms, whether the medicine has any effect at all, what the original disease was, or even whether the pills are having unanticipated side effects and are therefore responsible for the illness. All one knows for certain is that the patient is gradually worsening.

In any case, it is difficult to prescribe cures without a reasonable definition of health. We occasionally have had committees in the United States that articulated a long-term consensus about societal goals, then proffered a set of principles for the schools to employ in meeting those goals — stating what students should learn and how schools should assure that learning. Such committees were particularly influential in the late nineteenth and early twentieth centuries.

However, 'decisions' about fundamental principles today (and perhaps then, too, more than we suspect) are usually made by indirection, based mostly on sensed, rather than stated, popular sentiment — with periodic reversals when the mood changes. No national crisis of the magnitude faced by Japan and Germany upon the collapse of their governments after World War II has ever forced Americans to reassess the educational enterprise as a whole, and take the initiative to create a new system to meet new problems. Instead, the system was born in the exigencies of the nineteenth-century culture and classroom, and has been patched ever since, as needed.

As we learned more about high schools in the Stanford neighborhood, we were struck by what we have come to believe is a fact: while the system does not seem to show much intentional thought over the decades to achieve consistent, educational aims, a degree of valuable stability seems present nevertheless. The common schools in the nineteenth century, that great and uniquely American creation designed to serve all the children of an increasingly diverse population, is still trying — sometimes articulately and effectively, and sometimes not — to do the same thing. It has been adapted, some might say transformed, from what it was more than a hundred years ago; we do not have as many one-room schoolhouses, and the young people in public schools are much older than they were then. But the aim, in the broadest sense, is the same. That laudable intent of serving everyone is an overriding element in trying to understand the problems these institutions, along with their supporters, face.

Everyone Is Served; Everyone Helps Decide

In the early, formative years of this system, according to Kirst, 'Experience drawn from the testing of a jumble of ideas, ideas transmitted through new professional journals and new training for the emergent profession, did far more than the political system...to impose a striking uniformity on American instructional practices'. There was a resulting inner consistency, more or less. This system was expanded over many decades to include more people, for longer periods of time; it has incorporated the demands of broad social movements, shifting demographics and changing modes of production. And much of the resulting change in schools, Kirst reminds us, was stimulated outside the formal political structure of the country.

Now, we use political systems more overtly and systematically than we ever did in the past to effect changes in the schools. However, and partly as a result, the schools we saw have become even more markedly a group of institutions reflecting both subtle and obvious compromises, designed according to no grand plan. There is no reason to believe schools elsewhere in the United

States are any different. They serve best no one group exclusively — not the students collectively, nor the staff, nor any one income or ethnic group. To many educators, and to the public, the educational system seems jerry-built — constructed by accretion and, from any one perspective, distorted. As one teacher noted wryly:

> In the early 1960s, we lived in the shadow of Sputnik, with its attendant focus on padding the science curriculum. Then, just as we got our curriculum in shape, we came under the domain of the affective, the Esalen-type, researchers who were more concerned with feelings and potential than with technology. We've learned to endure the swings of the pendulum as it goes back and forth. A pendulum, by definition, is going nowhere.

The notion of a comprehensive high school, which, in the last analysis was enshrined by formal political action at various levels, has been one of the deepest, longest-held ideas in our nation. But perhaps that concept is being shunted aside, not consciously and not with intent to change the broader goal, but because it doesn't seem to produce the results everyone wants. In a paper titled 'The Comprehensive High School and Its Curriculum', one researcher in the Study, John Agnew, discussed American faith in the comprehensive high school:

> It has been argued that the so-called failures of the secondary school which have been reported at some length in recent times are in large part due to the fact that many of [educator James] Conant's specific recommendations went unheeded. An alternative view, however, is that the comprehensive high school ideology is fundamentally flawed. In particular, while its altruistic ideals are unarguable, it purports a practical mission for a single institution which is not possible given the current physical structure of the high school — as a building, and as a set of organized and structured classes. The American high school is an idea as much as a real institution.

Maybe Agnew is correct. The role the high school is expected to fulfill in the protection, projection and development of the American dream is remarkably diverse, and this role has widened, along with the social and education needs of American teenagers, in the twentieth century. But it is important, we believe, for Nation-building as well as for 'education', to recognize the power of the rhetoric of comprehensive education, and of the ideals behind the label. In instituting changes in the schools, it seems to us and to those who worked with us in the Study that we must take care not to modify the goals, even if we modify the workings of the institution itself. One of the most heartening impressions that emerged from the Study for us was the unmistakable commitment by teachers and school administrators to serving all the children and, further, to help them overcome barriers of class, race and abilities.

Schools have been subjected intensively to the potent pressures of interest-group politics in recent decades, as have many other aspects of American life. The result, however, is generally consistent with the history of education in America: to try to serve everyone. In the past twenty-five years, that means more of the poor, more minorities, more of the handicapped, more who do not speak English as a first language, more delinquents (since schools do not have the same powers they once had to expel children), and more older teenagers. Over the long sweep, that momentum toward inclusiveness is the great success story and problem of American public education.

The rapidity of change in the last two decades also has produced many of the problems of questionable and unequal quality that now have commanded national attention, and that we saw in nearby schools. Now, however, it will be no one person's responsibility to decide the future of American public secondary education, but everyone's. That is how educational decisions are made these days.

American educational politics are fluid and passionate. People care, lots of them. The cross-currents are swift and shifting. In such circumstances, we have come to believe strongly that the professionals — the teachers and school administrators — must become the centers around which public action rotates. Much of our orientation and the heart of our argument are based on this conviction: it is absolutely necessary today to find, first, the foundation for steps to fortify the professionals actually working in schools and school districts and, second, to make it easier and more natural for them to enter discussions with segments of the public, and particularly with authorities at the state level, from positions of strength.

Seeds of Regeneration

At some early meetings in the Study, as we have reported, Professor Kirst approached the participating high schools with a research plan designed to find major alternatives to the comprehensive high schools — what he called 'radical surgery'. However, when he went to the schools to court their partnership, he found that '*none* of those people were interested in that kind of a study. They didn't see this kind of alternative having anything to do with their work in the foreseeable future', he recalls.

Because the Study was collaborative, it tied itself by design to the relatively short-term interests of the schools. One consequence was that we were conservative in our perspectives and in our actions; as we have said, we self-consciously took an incrementalist view of educational change. 'A retreat in Aspen for academics and policy types would have come out with very different results. . . . We went out of the business of radical alternatives, and began to look at improving the existing structure,' rather than continuing 'visionary, utopian work', we were reminded later by Kirst. We recognize the fact that the approach we took may help perpetuate a succession of short-term, patch-up solutions to fix a system that may need fundamental restructuring.

Dornbusch encountered the same dilemma when he approached the schools to develop a research agenda. The idea that interested them was not long-term re-examination of schooling, but how today's families affect their own children's education. It proved to be a fruitful idea, and a jewel within the Study — but it did not and will not change the fragmented nature of education, give us alternatives to the comprehensive high school, nor even overhaul the tracking system that separates the college-bound from the vocational student.

In the current spate of educational reforms, in fact, few have promulgated a philosophy of what schools can and should teach, and what they should jettison. Theodore Sizer, with *Horace's Compromise*,[2] is a notable and controversial exception. But even Sizer's impressive contribution to our understanding of schools and his prescriptions for the future do not represent a consensus; rather they are the reasoned judgments of one observer, albeit a scholarly and wise one, working with those of like mind and conviction. As we have said, it is very likely that without at least implicit agreement on aims, general changes in the schools will continue to be disjointed, now serving one audience or meeting one problem, now another.

Perhaps such an outcome is inevitable; some may argue that it is desirable. Variation in education, as with living species, may be necessary for survival. We ourselves are not particularly fond of detailed, educational planning — preferring to believe that visions of the future must continually be tested against shifting values and unforeseen circumstances — illuminated by attractive variations; every educational plan, however wisely and carefully developed, must be modified regularly, even frequently.

To a significant degree, we already have a critically important base upon which to build, one that emerged clearly during the Study: the dedication, energy, and enthusiasm that still exists in the high schools — despite years of public castigation and denigration, and despite an almost complete lack of opportunity for professional revitalization.

Professor Dornbusch, for example, constantly asserted that *all* six of the principals he worked with were 'dedicated to education, hard-working and smart'. All were deeply in touch with today's children. Though they may not have all the answers, they understand the questions.

In the Study, we could not call extraordinary levels of professional commitment by adminstrators and teachers unusual. Today's schools are characterized, in fact, by dedication. (The cynics in the Study were fond of asserting that there is no other reason than personal dedication to stay in the profession, given the low pay and low levels of support.) And we found repeatedly that, when given the opportunity to reflect and be creative, local educators were ingenious in finding their own solutions to local problems.

We believe firmly that schools contain within themselves the seeds of their own regeneration. But the key lies in knowing how to identify those seeds and then nurture them. The imagery of 'nurturing' those seeds of regeneration is intended to help dispel the expectation of a quick fix. Nurturing, in this context, implies: more reflective time for teachers and school administrators, the autonomy that helps to translate the results of that reflection into better educational practices, the possibility of teachers working with fewer students, and more support services for teachers and children — among many other things.

Almost all of that means more money. In education, we are convinced, one can get more for less only at the margins; it is impossible with the core functions of schooling, at least not at the level at which the schools of California are funded at the present time.

Typically, the public underestimates the complexity of undertaking educational reform and its attendant costs, and is impatient with its slowness. But the kind of regeneration we envisage will require extensive legislative reform — even pruning back much of the overly-restrictive laws and regulations — and this will take time and resources at the State level. Figuring out how funds can best be allocated to improve teacher education, pre-service and in-service; to overhaul the school plant; to raise teacher salaries; to enhance curriculum revision — and much more, besides — means years of work, lots of goodwill and an unusual amount of patience. While schools have the capacity to regenerate themselves, we believe, they will not succeed in meeting the demands they face without plenty of help.

And these steps will take us only part of the way. After we reach a satisfactory base level, it then will be necessary to figure out how to create for schools, and the people who work within them, the necessary time to attend to local problems, generate local responses and sustain attractive local variation — all while trying to balance these efforts with a vision of what Americans want from their schools in the society at large.

We came away from the first few years of the collaborative work believing that a public commitment to the schools is needed over the long haul. If we had a societal consensus about what education should try to do, then, when a new wave of demands rises, we would have a steady compass by which to steer. But we think it unlikely that a clear and articulated consensus about goals will appear soon, or ever. In its absence, the only reasonable substitute that is apparent to us will have to be a new level of confidence in teachers and the school administrators. That confidence will have to be earned, but we believe the foundation does indeed exist for that to happen. Now we have to build on it.

Notes

1 Stanford Alumni Association, 1984.
2 THEODORE SIZER, *Horace's Compromise*, Houghton Miflin, 1984.

Chapter 12

Some Personal Reflections about the Collaboration and Its Future

Two of us, Kennedy and Atkin, took the lead in initiating the Study from bases in our respective positions of President and Dean of the School of Education at Stanford University. Both of us have long-standing and deep interests in education at pre-college levels; however, we also were committed to using whatever influence we had by virtue of our administrative positions to promote changes at our own University. We wanted Stanford as a total institution to become more active in improving educational quality, and modify some of its ways of doing business in the process.

Insofar as Stanford projects a general picture of its own role in the improvement of secondary-school education, to itself and to others, we wanted to see how far the institution could move from that role being reflected primarily by a collection of relatively disengaged and discrete commentaries on educational problems in secondary schools, to one of significant and somewhat coordinated involvement in grappling with those problems directly. And we thought that a route to that goal was to foster closer intellectual exchanges between the Stanford community and those in nearby high schools.

Both of us were experienced enough to know that institutional changes are not easy. They may be particularly difficult at universities, in fact, especially at the good ones, because traditions of professorial independence are strongest there. Nevertheless, we thought we might have some effect, at least at the margins, both through what we and others did in connection with the Study, and also by what we said publicly about it.

We had reason to think we might make progress. Almost every professor cares about the quality of education, and many of them are passionate about it. An overwhelming number of university-based scholars, however removed they may be in their daily lives from the practical implications of their work, believe that ideas have consequences, particularly their own ideas; and they harbor ambitions, not always entirely suppressed, to have a major impact on some corner of the practical world.

This outlook on the relationship between scholarship and human affairs characterizes professors in general, but it is especially prominent among faculty in professional schools. Regardless of the degree to which professors in such schools strive for higher academic standing through the generation of scientific theory, they know that their long-term institutional existence depends on other people noting significant links between scholarship and teaching, on the one hand, and the improvement of some professional practice, some aspect of the human condition, on the other.

We felt also that teachers and school administrators would welcome the chance for closer association with Stanford. Mostly, they would value the chance for reflection and intellectual stimulation that would be possible if blocks of time regularly were set aside to talk about their professional activities, and if professors and graduate students came to those discussions with first-hand knowledge of life in schools. And we knew that Stanford, in general, was well-regarded among school people in the broader community. Our neighbors in the schools had worked for many years with our student teachers, and were impressed; many of the student teachers, on graduation, had been sought for regular teaching appointments in the very districts we intended to work with. We were fortunate, furthermore, that the many Stanford professors who had done research in the nearby schools had left favorable impressions. (This last point may strike some readers as unusual. It is.)

We sensed also that school administrators, in particular, saw political benefits to be derived from closer association with the University. Not only might an institution with more prestige and prominence than their own help legitimize their practices, the University could serve on occasion as a bit of a buffer between their in-many-ways embattled institutions and an often critical public.

Finally, we believed that there was an emerging climate of support for our initiative in the broader community, as politicians and other influential figures seemed to be turning seriously to issues of improving pre-college education.

Establishing close associations between the two educational sub-cultures of high school and university, however, is far from straightforward, as we have tried to indicate in our narrative. Expectations of the various parties in a new venture are often notably different one from another, and sometimes in conflict. It is only slight oversimplification to state that some professors considered work in the schools to be collaborative if teachers gave consent for questionnaires to be administered to their students — and some teachers hoped for detailed prescriptions from the University to cure all the school-based ailments that they identified as most important.

But these polar views were not usually prominent during the Study and, to the extent that they existed, did not seriously impede the work. Mostly, the two institutions accommodated one another, though not always smoothly. The specifics of the collaborative process ranged broadly. In some cases, there was strong participation by all parties every step of the way — from problem identification, through close working relationships in analyzing, interpreting and plumbing meaning from the data, to development of concrete steps to improve things. At the other extreme, we had to remind ourselves, in one instance, to share data with teachers and school administrators before information was released to the public!

When misunderstandings arose, as in the very painful and public case of our releasing certain sensitive findings (about the low priority put on significant intellectual activities in many classrooms) before full discussion of the meaning and implications of those findings with teachers and school administrators, the problem proved manageable, in the sense that it did not establish an insurmountable obstacle to sustaining the ongoing collaborative activities. We found we could tap a deep reservoir of goodwill among school people toward Stanford and the Study, to which we alluded earlier. Partly through the sensitive and prompt reactions of the professor whose work had become known prematurely (and partly through the damage-control efforts of the co-Directors), we were able to overcome the few voices, like one high-school principal's, that urged a retreat from the Study because of our serious mistake.

Some Disappointments

It is not for us to declare the Study a 'success', though later in this chapter we outline some of our reasons for feeling a sense of accomplishment for ourselves, for our Stanford colleagues and for those in the schools with whom we worked. But as we look back on the initial years of the collaboration, the years that encompass the activities included in this report, we find ourselves still disappointed in three areas of importance. We elaborate here on each.

First, in only a few cases did we achieve the intensity of collaboration to which we aspired. While we knew that the problem-identification phase of each project was to be largely the work of Stanford-based people, and we were able to articulate what was for us and also for those in the schools at that time an acceptable rationale for such an approach, we envisioned a process thereafter whereby there was systematic and significant collaboration. We even idealized a set of 'steps' through which Stanford researchers and those from the schools would progress:

1 Does the information that has been collected correspond with the experience of teachers and school administrators in the district? Do the data ring true?
2 Does the situation highlighted by the data reflect a condition that should be changed? That is, do the data illuminate problems that those in the schools wish to address?
3 How do district faculty and adminstrators account for the findings? What circumstances have led to the current condition?
4 Knowing some of the factors responsible for the condition, what lines of action, if any, seem appropriate in light of the findings?

We thought that as discussions proceeded through these stages, we would move progressively to styles of thought and modes of action more congenial to those in the schools than to professors, though we hoped and expected that those from Stanford would continue to make some contributions, right down to the implementation of new practices. In only a few cases, however, did this degree of collaboration actually occur. Where it did happen, most visibly with Professors Dornbusch and Garet, the results were powerful and gratifying.

To be sure, there was plenty of collaboration in the activities spearheaded by professors others than these two, even if it did not precisely fit the model we had in mind or reach the intensity we hoped for. In Professor Weiler's case, in fact (and he was the professor who worked most closely with teachers, as contrasted with superintendents and principals), there was heavy involvement by those from the schools in identifying the researchable issues in the first place. In other instances, professors systematically obtained papers, or reactions to papers, from teachers and administrators. In several cases, they helped to design the instruments for collecting data, as with Professor Dornbusch — and even then collected them, as in some of the shadowing studies supervised by Professor Eisner. But, still, we came away from the initial, three-year period wishing there had been more intellectual exchange between those from Stanford and those from the schools.

While the concept of collaboration has a full set of attractions, not least are the substantive insights to be gained from those closest to the scene and, although we got a great deal, we did not manage to obtain as much of that as we wanted during the first three years, particularly from teachers.

The 'fault' is not theirs, however; almost certainly it is attributable to prevailing styles of intellectual life at the University. Very few professors have internalized the fact that analysis of professional practice, *as it proceeds*, is an important source of knowledge. In professional activity (and in most other aspects of life), we sometimes do not fully understand until we act. A teacher may not perceive his or her own priorities, as one example, until he or she wrestles in the classroom with the question of how to deal with a particular child's apparently irrelevant questions. Academic norms, on the other hand, favor thought *prior to* action and reflection *following* action.

This troublesome and artificial division between theory and practice, between thought and action, has been understood since Aristotle wrote about it more than 2000 years ago. He, and many others since, have emphasized the moral dimensions of practical activities. A teacher, a physician, a lawyer and a social worker must inject moral considerations into their choices. Science helps, but it's not enough. University-based scholarship has not been particularly sensitive to this fact. True, professors in business schools, law schools, education schools and medical schools are beginning to appreciate the importance of understanding practical worlds to generate useful knowledge, but we still have a long way to go.

A second kind of disappointment is the fact that more scholars outside the School of Education, particularly in the curriculum fields that constitute the academic program of the high schools — English, history, biology, foreign languages, mathematics — did not become engaged in our work. To be sure some did, particularly in the project on American Schools and the World, but not in the numbers we had hoped.

At one stage we had a meeting with professors in the humanities who might be interested in more intensive work with high-school teachers in these subjects. There was also a session with science and mathematics professors. We had a Steering Committee composed of about ten professors and administrators at Stanford from outside the School of Education (in addition to several from inside the School) that met several times during the first three years to hear about research in progress and make suggestions. There was indeed some interest expressed in participation in the Study in each of these three groups, but we were not able to capitalize on it during the period about which we are now reporting.

On the other hand, we had significant involvement from social and behavioral scientists outside the School of Education. Professor Sanford Dornbusch of the Department of Sociology was one of our stars in his dedication and skill in working with school administrators. Professor Mark Lepper of the Department of Psychology engaged himself wholeheartedly in the work of the technology project, and chaired it at one stage when Professor Walker was away from campus. Ron Herring and David Grossman of the Center for Research on International Studies played central roles in the investigations of education in global affairs in the schools. Each of them, and several others, did pioneering work that was much acclaimed during the course of the Study. But we had hoped for more work than we were able to generate on subject-by-subject curriculum examination and revision.

This circumstance, in retrospect, might have been predictable. The School of Education, during the 1960s and 1970s, had conscientiously staffed itself with outstanding social and behavioral scientists. However, in many instances, these scholars were brought to the University to replace professors with subject-specific interests in curriculum matters. When Paul Hurd, Stanford's renowned specialist in science education, retired, he eventually was replaced by a social scientist. When Ed Begle, an internationally respected mathematics education specialist, died, he, too, was not replaced by someone with similar scholarly interests. This kind of substitution in specialization was not uncommon in the School at that time. It was undertaken, in part, to strengthen the research capability of the institution, and, in part, to underscore the importance of new and important perspectives on education that were surfacing among social and behavioral scientists. The move succeeded, impressively, in what it set out to accomplish — but at the expense of weakening the School in fields of basic professional competence that were to loom of high priority a decade later.

As a result of the new staffing pattern of the 1960s and 1970s, however, the School strengthened its links across campus with fields like economics, anthropology, linguistics, sociology and political science, and the Study could, and did, take full advantage of those connections.

A third area of disappointment for us was the fact that Stanford, but particularly the School of Education, did not directly and immediately see the Study as a base from which to examine its own programs. We had envisioned a study of Stanford *and* the schools, not solely a Stanford research project *on* the schools, however collaborative it might be. Implicitly, it turned out, we were indeed re-examining our scholarly styles, and trying something new. Inevitably, our teaching as well as our research was changed by the effort. When professors engage in new activity, there is quick translation to the classroom, particularly at the graduate level (and the School of Education is a graduate school). Even more potently, literally dozens of graduate students, inside and outside the School, were engaged in research that was part of the Study, and their scholarly outlook — and their work for decades — was influenced by the styles of collaborative research with people in the schools they were helping to pioneer.

But we were looking for a more self-conscious examination of programs at Stanford — particularly, as one important possibility, a re-structuring of the teacher education program. Neither of us underestimates the power of a change in thinking on the activities one engages in, professors included, but we had hoped for something more visible. In all probability, we were motivated by one feature of our own ideology about the Study: we thought it presumptuous for one institution to be studying another for purposes of making changes; a more appealing stance was a study of both institutions, cooperatively. The fact that it did not occur more explicitly, for Stanford, is a source of dissatisfaction (though, as we write, and with the benefit of our knowledge of some of the follow-on activities initiated as a result of the Study, our feelings of discomfort are somewhat eased; our teacher education programs are being re-evaluated as part of the new record of collaboration that succeeded the initial three-year period, and in loose association with a pioneering study of possible new examinations for teachers that has been launched on campus).

In retrospect, we probably would have had greater success in directing Study-sponsored activity toward needed changes at Stanford if we had been more explicit about it, and if we had established sub-groups to consider the matter. In fact, we did make effort in that direction by establishing a teacher education project. That effort, however, never progressed during the first years to direct assessment of Stanford's own program. Perhaps the issue was too sensitive for us. Perhaps we were not ready then. Still, we probably should have worked harder at it. At the very least, such activity would have signalled to our collaborators in the schools that Stanford was willing to take some risks, too.

We have tried to detail and elaborate here the fact that we fell short of our own hopes, but one fact and one perspective that was encouraged by people we respect softens our view of our lack of complete success in the three realms mentioned above.

First, the fact: though we did hear some harsh complaints, there were actually very few expressions of disappointment from superintendents, principals or teachers about the level of collaboration. There may be many reasons: perhaps there was, as a direct result of the Study, more contact with Stanford than there had been, and that was valued; or perhaps the energy and commitment of professors and graduate students overcame any feelings of unmet expectations; or perhaps those in the schools expected less from Stanford than we did, taking our initial declarations with more than a few grains of salt; or perhaps Stanford was seen as sincerely searching for the positive side of school practices; or perhaps a combination of all these perspectives — plus others. But, whatever the reason, we were accepted in good faith, and evidently the failings detected by the three of us were not seen as severe enough to threaten the new relationships seriously.

There was considerable evidence to the contrary, in fact. We continually received expressions of support and encouragement from those in the schools with whom we worked. A particularly thoughtful and gratifying assessment was submitted near the end of our initial period of work by Joyce Rosenstiel, Principal of Menlo-Atherton High School:

> Sequoia District entered the Stanford and the Schools Study in its second year, choosing two of its schools as participants. I looked upon my involvement with quiet skepticism. As a woman principal, I had often been 'researched' by graduate students, questioned, shadowed, asked to evaluate my day in percentages of administrative versus instructional time. Often the pursuer of advanced degrees didn't give even the courtesy of a copy of finished materials, reinforcing my belief that university-based research had little or no impact on practitioners.
>
> In fact, agendas of university staff and practitioners are often *not* congruent, as researchers are under pressure to publish as a way of furthering their careers, while practitioners function in the short range, making daily decisions based as often on intuition as research data. It was in this frame of mind that I sat through the meeting where the Dornbusch research design was first proposed.
>
> Something struck a deep chord as the hypothesis was explained. The idea of exploring family process to see if successful family patternings could be determined and made available to all families, promised to open exciting doors. Family structure is a given entity

that school cannot change. Would participation in the study throw light on current family patterns, updating what we had observed and had hunches about, by adding statistically accurate information?

Intrigued by the idea that families of the 80s could be given solid information that would promote self-change. I joined other principals in pledging time and energy to the project.

I have not been disappointed by my decision. The best qualities of this project have been the true collaborative nature of it from the beginning. Principals felt completely free to dissent, to shape survey questions, to indicate directions they felt would lead us to our goal.

In addition, there have already been concrete results for all of the participants — parents, school personnel and university researchers. The information gained from the questionnaires has been widely distributed to families, in newsletters and parent meetings. The nature of schools being what they are, the data must be given out again during this school year as 50 per cent of the student body is new. It should, however, be more persuasive because the results were gathered locally at our own schools.

Other intriguing possibilities suggested by the research remain to be pursued. There is the possibility of reaching selected constituencies, to help them 'reverse the odds'. For example, given the information revealed by the data analysis, that too early autonomy for adolescents negatively impacts grades, could single parents be set up in support groups to share ways of becoming 'authoritative' in interacting with their children, hoping to see direct results in better schoolwork? What could the school's role be in working with such selective groups?

Finally, there have been unexpected outcomes. Although the principals initially responded and participated because they were intrigued with the hypothesis, all of us have developed close professional ties while working together. This has promoted cooperation in areas not related to the Study. As a direct result of friendships formed within the group, the San Mateo district 'loaned' a librarian for a semester to help cover our sabbatical need. Other problems can often be solved by picking up the phone, knowing that there is a cooperative problem-solver on the other end.

Our work together suggests that collaboration is not only possible but productive for all of the participants. It has also suggested other areas where cooperative research could provide answers that would result in direct action at school sites. Many hours have gone into the work, but it has been worth it.

Perhaps the degree of collaboration that we idealized and to which we aspired, the novel, inter-institutional associations we hoped to establish on a large scale, inevitably takes time. Joyce Rosenstiel's thoughts on the matter that are reproduced here did not become apparent to us until she read an early draft of this book.

Three years simply may not be a sufficient period to achieve the kinds of relationships we hoped to see. We were, after all, not talking about an add-on activity to Stanford or the schools, a marginal 'center' or 'institute' to demonstrate our social conscience, staffed essentially by temporary people brought on only for the life of the Study and the period of sharpest societal concern. We were trying instead to modify the core activities of a significant number of professors — in effect, were trying to shift somewhat the nature of their scholarship, not just for the duration of the Study, but for the indefinite future.

This view of a monumental goal that takes many years to reach was signalled early, vigorously and persistently by our own National Advisory Board, which consisted of Norman Francis, John Gardner, Henry Gauthier, Patricia Graham, Fred Hechinger, Shirley Hufstedler, Frank Macchiarola, Robert Maynard, David Saxon, Albert Shanker, Paul Simon, Rocco Siciliano and Dean Watkins. The issue arose prominently at the first meeting of the group and was to resurface regularly, in one form or other, at every other meeting. In the initial version, at our very first session, we at Stanford had raised the topic of publications that might arise from the Study, as professors are inclined to do. The topic was not taken up with enthusiasm by our advisors.

First, said they, drop all plans for a 'final report'. If academic norms or funding requirements propel Stanford toward some kind of publication, find another label. The essence of the Study, the appeal as far as the Board was concerned, lies in the process of collaboration, which is novel and difficult. To release a 'final' report would send exactly the wrong signal: the University entered the schools for a few years, told the world about what they saw, then moved on to other things. The high schools need steady and persistent attention from the higher education community.

Furthermore, the relationships already in evidence between Stanford and people in the schools, to the Board, seemed a significant beginning. We were told that the kinds of collaboration we had started take a long time to take shape, that we should expect disappointments, but that the key lay in persistence. We were, after all, engaged in an ongoing cooperative research enterprise, not a familiar activity in which we provided solely a service with which we already had a great deal of experience, like teaching classes to administrators or teachers. The level and depth of the intellectual relationships we were trying to foster are not achieved quickly, even apart from the circumstances of the sharply different sub-cultures we were trying to bring together.

Continuing the Collaboration

As the initial funding period for the Study was drawing to a close, we began to plan for continuing the collaboration between Stanford and the nearby schools, a key goal in initiating the Study in the first place.

As a first step, we asked the superintendents who had been working with us their views about continuation of collaborative work, and they expressed unqualified commitment to and enthusiasm for extending and expanding the associations. There was no dissent, not even a reservation; at an early meeting, they even stated that they were willing to contribute a modest amount to the effort financially, if necessary, because the Study was clearly beneficial — to them, to their principals and to their teachers. In today's budget climate for California schools, there can be no stronger commitment. Mostly, they said, they valued the opportunities for the intellectual exchange.

At the same time, we sensed significant interest at Stanford in continuing collaboration. Some professors, like Dornbusch and Eisner, had already started new research programs based on their work in the Study that they wanted to continue. Additionally, several professors who had not been part of the effort during the initial three-year phase of collaborative activities stated that they now wanted to participate. Apparently, they were coming to understand how their own scholarly activities, and their teaching, could thereby be improved.

Of prime importance, Professor Larry Cuban, who soon was to be designated the new Associate Dean for Academic Affairs of the School of Education, and who had not been involved in the first phase, said that he would be willing to assume administrative leadership for designing the follow-on set of activities to the Study of Stanford and the Schools. Cuban had joined the faculty in the early 1980s, coming to the University directly from seven years as Superintendent of Schools in Arlington, Virginia; he was committed to strengthening the links between Stanford and the nearby schools, and soon after joining the faculty had organized a well-regarded 'Superintendents' Roundtable' that had been in operation for several years. No one in the School of Education was better suited than he to carry the work forward.

Cuban soon began meeting with teachers, school administrators and Stanford professors to plan for continuing cooperative activities — but designing the next phase to profit from the first. It is not our purpose here to describe the new activities in detail, indeed they are just beginning as we write, but let us note that some key changes were made: people from the schools were to play a stronger role in identifying the issues that would be studied, class-room teachers would be more active than before, the work would be expanded to include elementary schools, additional schools districts would join the initi-ating six, the name of the effort would be changed to the 'Stanford/Schools Collaborative', and more.

Thus the continuation of the activity was assured; a major goal in launching the Study was achieved. Apparently, we had demonstrated to a large, busy and independent group of people, those in the schools and those at Stanford, that there were sufficient benefits in a research-based collaboration between schools and a nearby university to continue a novel and difficult enterprise.

Looking to the Future

Thus we created a study that turned out to have a life beyond its own. But what, exactly, will that future life amount to? Will it generate a real change in the schools and their teachers, and in Stanford University and its students? Or will it add up merely to an extension, in a few interesting areas, of the subject matter of the Study itself?

Obviously that will depend in part on how energetic we and our successors are able to be. It is always a good idea, however, to announce one's hopes; we do so cautiously, and with full respect for the natural inertia of large and complex institutions. At the same time, we do so with great regard for the importance of improving higher education's leadership of the great national venture of schooling — not only for the needs of the pre-collegiate insti-tutions, but for those of the universities and colleges as well.

We hope, first of all, that the School of Education will be able to continue and to extend its concern with the practice, as opposed solely to the *theory*, of education in the public primary and secondary schools. The history of such schools in distinguished research universities has been one of ebb and flow, toward and then away from contact with the profession. Stanford's own historic oscillation is recorded briefly earlier in this report; Harvard's was chronicled by Derek Bok in his 1985 *Report to the Board of Overseers of Harvard College*. The similarity in alternating approach and avoidance of direct involvement with the schools suggests a fundamental ambivalence about mission, one perhaps related to the short history of these professional schools and their difficulty in gaining acceptance as full partners in the academic enterprise. As policy-related research becomes more accepted and as the schools become more clearly recognized as important and appropriate subjects for analysis, that may change. If it will not change by itself, then we need to make it change: we can ill afford the survival of a situation in which the very activity on which our universities depend most critically, indeed the activity most closely related to the one they perform, is an object of lack of interest, or even scorn, within them.

For that to happen, there needs to be significant expansion of the involvement of faculty outside Education with the schools and what goes on in the schools. Here, we suspect, there is a difficult problem — but one that needs to be faced with candor. Education as a process is not much inquired about even within, that is at the level of, the university. Improvement of teaching, examination of how learning takes place, experimentation with new methods — all these are things that happen in universities, but they are hardly matters that achieve broad prominence. Rather, there tends to be a program here, an interested department or faculty member there, but seldom a well-mounted and well-subscribed institutional effort. Thus it is hardly surprising when faculty are difficult to recruit for the tasks of involvement with the schools. Nor is it difficult to understand why educational leaders from the pre-collegiate sector are skeptical of proffered help from the university level.

Nevertheless, there are signal examples of successful involvement of university teaching faculties with local school districts. Some of the most successful have occurred at universities without schools of education, suggesting that perhaps where such a school is present other faculty members may be happy to 'leave it to the Ed School'. But there is no reason why a school of education should not be able to function as an effective catalyst in involving other faculties, and we hope that in the future Stanford may provide an example for others to follow.

If the difficulties can be overcome, the gains to be realized are great. We learned in the great curriculum reform movements of the 1950s and 1960s how important it was to secondary-school science teachers to be swept into membership in a professional community. Through National Science Foundation summer institutes, through visitation programs, through joint textbook-writing efforts, and through various in-service training programs, university scientists and high-school science teachers were thrown together in collegial fashion. The university people learned much about the challenge of effective teaching; the secondary school people took away some of the excitement of new results and new technologies. Most of all, there was a gain in respect and understanding — one that, for a while, made a significant impact on the morale and enthusiasm of the high-school teachers. Without substantial Federal support of that same kind (for that to materialize, international competitiveness may have to become the moral equivalent of Sputnik!), that kind of continuity can perhaps only be established by massive university outreach into the pre-collegiate system. It could produce the largest effect on the status of the teaching profession of any intervention we can conceive.

What would the universities gain? First, of course, a resolution of the dilemma of the schools of education. Transformed into leaders of a substantial outreach effort, they would gather strength and influence within their own institutions, and gain access to a host of new research challenges. Outreach is possible without them, of course. But where schools of education are present, those faculty members should become the confident proprietors of a successful and creative relationship with an increasingly important profession — and also the mediators of a vigorous research commerce.

Second, the universities could have a dramatic impact on the views and the commitments of their own students. On many campuses, the voice of obligation — as a counterpoint to those of personal success and entre-preneurism that one hears on all sides — is much too small. The theme of public service is emerging in many places; where it is strongest, the major chords are preparation for employment in the public sector (through programs emphasizing government internships and the like) and part-time voluntarism. And programs in the latter category involve, overwhelmingly, a single activity: *teaching*. Whether it is tutoring educationally disadvantaged students in an inner-city school, working in a national program for adult literacy, or reading to the elderly blind in a senior citizens' center, the volunteer activity of a college undergraduate is highly likely to entail work that is like the teacher's work. Thus there is a prospectively powerful synergism between the movement of the universities toward the schools and the movement of their student bodies toward public service. Formal involvement at the institutional level with school improvement can provide outlets for the participation of increasing numbers of students, both undergraduate and graduate, in research on education and in teaching as well. Conversely, visible student engagement with, and enthusiasm for, the business of teaching will provide added reasons for undertaking some change in the universities' direction.

Finally, we believe that all these related opportunities require, and in turn will stimulate, enhanced participation on the part of universities and their leaders as active spokespersons for educational reform. Consider what has already gone on. In 1983, the presidents or chancellors and deans of education from half a dozen influential research universities met at Pajaro Dunes in California to discuss ways in which they might help to provide more impetus for helpful reforms in public pre-collegiate education. All of those participants have since been active in drawing their own institutions closer to local schools and/or in advocating more national attention to school problems. There was conspicuous participation by university leaders in the Carnegie Forum on Education and the Economy; as a result of its excellent report on the status of the teaching profession, *A Nation Prepared*, a meeting of presidents from various institutions of higher education considered how to gain more attention and national support for its major recommendations. That has been followed, in September 1987, by a meeting near Minneapolis of over forty presidents who agreed to try and fulfill that purpose. It is not entirely accidental that the membership of that group overlaps strongly with that of Campus Compact, a consortium of over 200 institutions committed to the active support and extension of public service programs on their campuses.

These efforts offer the prospect of major change in the attitude of higher education in the United States toward a family of public obligations, among which the health of our educational system is the most central priority. After all, nothing is more directly connected to the mission or the self-interest of the universities than that. Schools of education arose in the first place as devices for assuring that properly-prepared students were available in adequate numbers for entrance to the universities that established them; and teaching is, if not the only obligation of our colleges and universities to the larger society, at least *primus inter pares* in the family of those obligations. So undertaking that responsibility would be a natural extension of our tradition. What would it accomplish?

A renewed acceptance by the universities of leadership for the entire venture of education would strengthen the schools of education, and would provide new energy for involving faculties outside those schools in research and outreach programs. Secondarily, it would lend enthusiasm and vitality to efforts to analyze and improve teaching in the university itself; it is impossible to grapple realistically with that problem in the other sector without recognizing it as a problem in your own. And finally, it would support and extend what we believe to be a revolution in the making in the attitude of our best young people toward education in particular and their obligation to others in general.

These outcomes are, it may be argued, a long stretch from the limited and local success we have had with the research collaboration called the Study of Stanford and the Schools. But our experience would suggest that the connection may not be so very far-fetched. Successful and continuing involvement with local schools *is* possible. It *does* make it possible for the institutional leadership then to speak more actively and persuasively on behalf of broader political and economic support for educational intervention, and to undertake initiatives with other parts of the higher education sector. It *has* helped improve the capacity of the Stanford School of Education to run its teacher-training programs more successfully, and with better students. It *has*, if only marginally, attracted other Stanford faculty to work with the schools and improved the image of the School of Education in other parts of the institution. And it *has* been accompanied by a conspicuous rise in the interest of the best Stanford undergraduates in public service, including teaching.

So although we cannot claim to have found a future, we do think we have a glimpse of some bright prospects — so bright, indeed, that they could change the fate of the next successor generation if only we could make them happen. Research collaboration with the schools will not do that. What it did do, in our experience, was to start something — an unusual process through which the method and the collaboration itself became the subject of inquiry. In turn other things became possible; and eventually the central accomplishment of the effort bcame a family of events that were sequels to the reunion between Stanford and the schools. Perhaps we are hopeless optimists; but in those events we see the prospect for recovering our respect for schools and schooling, re-establishing the unity of the teaching profession, and regaining the commitment of our best young people to the activity that conserves and extends our culture. There is no guarantee that these things will result from university collaboration with the schools; but if it does not begin there, then what *will* get us started?

Participants in the Study of Stanford and the Schools

David Abernethy
John Agnew
Robert Alioto
Marilyn Baldauf
BankAmerica Foundation
    (John Alden, Caroline Boitano)
Lillian Barna
Bernadine Barr
Juliet Baxter
Charles Benjamin
Jane Boston
Liora Bresler
Alta Brewer
Christy Bergin
Mary Brouillard
Robert C. Calfee
Margaret Camarena
Albert M. Camarillo
Martin Carnoy
Lorna Catford
Steven H. Chaffee
Marilyn Chambliss
Community Foundation of Santa Clara
County
Lee Combs
Ramon Cortines
Lee J. Cronbach
Verdis Crockett
Brian DeLany
Dennis Devine
Sanford Dornbusch
The Educational Foundation of America
    (Richard P. and Sharon Ettinger)
Susan Duggan

Elliot Eisner
Nancy Ellis
Far West Laboratory
Shirley Feldman
Steven Ferrara
Angie Fischinger
Ed Flank
David Flinders
Martin Ford
Sandra Foster
Michael J. Fraleigh
Norman Francis
Ed Fried
Merle Fruehling
Ivan Fukumoto
Nathaniel Gage
John W. Gardner
Michael Garet
Henry Gauthier
Nicholas Gennaro
Roberto Gonzalez
Patricia Graham
Kathryn D. Gray
David Grossman
Sigrun Gudmundsdottir
Sarah Guri
Susan Hanson
Lois Harmon
Robert Harrington
Albert H. Hastorf
Edward Haertel
Rebecca Hawthorne
Fred Hechinger
Ronald B. Herring

Ralph Hester
The William and Flora Hewlett Foundation
(Theodore Lobman)
Robert Hill
Diane Hoffman
Shirley Hufstedler
The James Irvine Foundation
(Jean Parmelee)
Samuel Johnson
Robert Jurafsky
David Kennedy
Michael W. Kirst
Jean Kanerva
Joseph Khazzaka
Larry Kocher
Christina Konjevich
Meg Korpi
Helena Kraemer
Charlotte Krepismann
John D. Krumboltz
Vicky LaBoskey
David Lansdale
P. Herbert Leiderman
Mark Lepper
Hoover Liddell
The Lilly Endowment
(William Bonifield)
Thomas Lorch
Eleanor Maccoby
Robert Macaulay
Frank Macchiarola
Robert Madgic
Robert Maynard
Gary McHenry
John Mackay
Helen McKenna
John McManus
Richard P. Mesa
Milpitas Unified School District
Mountain View/Los Altos Union
High School District
Robin Munson
Worku Negash
Pete Newton
Charles W. Nichols
Nel Noddings
Mary Nur

The David and Lucile Packard Foundation
(Dolly Sacks)
William Paisley
Robert Palazzi
Scott Pearson
Claire L. Pelton
Charles Perotti
Denis Philips
Don Pope-Davis
Barbara Porro
Gary Poulos
Barbara Prescott
Philip L. Ritter
Donald F. Roberts
Sam Rodriguez
Everett Rogers
John Rogers
David Rogosa
Joyce Rosenstiel
Paul S. Sakamoto
Joel Samoff
Jesus Sanchez
Nancy Sanders
The San Francisco Foundation
(Bernice Brown)
San Francisco Unified School District
San Jose Unified School District
San Mateo Union High School District
David Saxon
W. Richard Scott
Aaron Seandel
Austin Sellery
Sequoia Union High School District
Sandra Schecter
Lisa Shaffer
Albert Shanker
Lee Shulman
Rocco Siciliano
Paul Simon
Derek Sleeman
Myrnalee Steen
Carol Leth Stone
Nancy Stone
Myra Strober
Pinchas Tamir
Carl Thoresen
Steve Thornton

Stephen Thorpe
Judith Torney-Purta
Caroline Turner
Angela Valenzuela
Marc Ventresca
Fox Vernon
Al Vidal
Bill Vine
Decker Walker
Dean Watkins
Hans Weiler
Kathryn R. Wentzel

Lindsay White
Sue Wiedenfeld
Cathy Wiehe
Jerri Willett
Gary Williamson
Choya Wilson
Marlene Wine
John Wirth
Ken D. Wood
Suk Ying Wong
Thea Zeeve

Index